# Pursuing Giraffe

## A 1950s Adventure

# Pursuing Giraffe
## *A 1950s Adventure*

ANNE INNIS DAGG

Wilfrid Laurier University Press

We acknowledge the support of the Canada Council for the Arts for our publishing program. We acknowledge the financial support of the Government of Canada through the Book Publishing Industry Development Program for our publishing activities.

ONTARIO ARTS COUNCIL
CONSEIL DES ARTS DE L'ONTARIO

Library and Archives Canada Cataloguing in Publication Data

Dagg, Anne Innis, 1933–
    Pursuing giraffe : a 1950s adventure / Anne Innis Dagg.

(Life writing series)
ISBN 978-0-88920-463-8 (paperback).—ISBN 978-0-88920-539-0 (PDF).—
ISBN 978-1-55458-662-2 (EPUB)

1. Dagg, Anne Innis, 1933–  —Travel—Africa.  2. Giraffe—Behavior—Africa.  3. Africa
—Description and travel.  4. Zoologists—Canada—Biography.  I. Title.  II. Series.

QL31.D34A3 2006                590'.92                C2005-907541-4

Cover design by P.J. Woodland. Cover photograph courtesy of Anne Dagg. Text design by Catharine Bonas-Taylor.

Every reasonable effort has been made to acquire permission for copyright material used in this text, and to acknowledge all such indebtedness accurately. Any errors and omissions called to the publisher's attention will be corrected in future printings.

Dedicated to the Memory of

Alexander Matthew
Griff Ewer
Jakes Ewer

who made my year with the giraffe not only possible
but profitable and indeed, a virtual paradise.

# CONTENTS

# FOREWORD

Anne Innis Dagg has written a brave and moving account of her time as a young white woman travelling and doing research in Southern and East Africa.

Mid-1950s Canada had not yet produced the personal or professional liberties feminist movements of the sixties and seventies would begin securing for women in North America. In a time when very few, if any, white woman scientists had engaged with Africa, Dagg's courage and determination forged the space for her to follow a childhood dream of studying giraffe in their African habitat. In doing so, she helped pave the way for later women to produce other work of enormous importance to science and conservation on that continent.

The two decades following Dagg's time in Africa would produce irrevocable changes in North American and African race politics. At the same time as the civil rights movement and demands for the rights of native peoples gained momentum in the Western hemisphere, Africa was rapidly loosening the grip of five hundred years of European invasion and occupation. Only eight years after her visit there, black people in East Africa won their political independence from Britain. In contrast to what was happening in East Africa, the time that Dagg spent in South Africa was the period in which the white government there was entrenching a vast range of racist legislation that would come to be known as apartheid. Dagg's experiences and observations as an outsider in South Africa offer provocative witness to the terrible ease and disturbing normalcy with which brutal political and racial systems are perpetuated by very ordinary people.

The young Anne Innis Dagg is a scientist. Academic. Woman. A little in love. Against the odds, she raises the money for her research and the journey.

She goes in humility, and with little of the stridency that characterizes so many of the narratives of those who went before or even after her. She goes on her own by ship. She is drawn to the oldest continent's landscapes, people, and animals. She will buy a car of her own. She will drive at night alone, only guessing at her own whereabouts. She will climb Kilimanjaro and be awed by the spectacle of the Victoria Falls. She will study giraffe in the wild and will end up grappling with what it means to be white in Africa. And so she will begin to confront her own discriminations, biases, and blind spots. She will celebrate democracy that will come—she knows—albeit almost four decades after she first went ashore in South Africa. At heart, the story is certainly about what she observes, studies, films, and chronicles in 1956 and 1957, details and understandings that had never before been documented about her subject: the beloved giraffe of Africa. But this book is also about one person's intellectual imagination, spirit of adventure, and daring: where she has long dreamed of going, where others either say she shouldn't or cannot go, and where some work against her going, she goes.

— Mark Behr

# PREFACE

As I was growing up in Toronto, my life's ambition was to go to Africa and study the giraffe there. In 1956, when I was twenty-three years old, my dream came true. This book is based on an extensive journal I kept during this adventure, the high point of my life, and on long, excited, twice-weekly letters I wrote home to Canada. It focuses first, of course, on giraffe, which I observed mainly on a twenty-thousand-acre ranch in South Africa near Kruger National Park, but also in Tanganyika, Kenya, and Southern Rhodesia. As a result, in part, of this work, I was able later to publish scientific papers not only on the behaviour of giraffe but also on their subspeciation, distribution, food preferences, gaits, and sexual differences in their skulls (Appendix 1).

This memoir also has two lesser areas of focus: the culture of racism and colonialism extant in the countries I visited at that time, and my own naïveté as a young woman educated at a private girls' school in Toronto.

The racism of most white people in the Africa of that day astonished me; I was unaware of such discrimination because I grew up in Toronto in the 1930s and 1940s, where there were few blacks or recent non-European immigrants, while aboriginal peoples lived largely out of sight on native reserves or in the North. Virtually everyone I knew was white. It did not occur to me that colonization in Africa was mirrored in the treatment of Indians and Inuit (then thought of as Eskimos) in Canada who, at that time, did not even have the vote. Nor did I realize that most Canadians were white because the country had a restrictive immigration policy. Canadians (unjustifiably) felt superior to Americans who, in the 1950s, were in the throes of trying to solve the problems of deeply entrenched racial discrimination.

Idealistic people such as myself were unable to comprehend the racism prevalent in Africa. How could it be that some people were denied schooling because of the colour of their skin? And then were derided because of their lack of education? I also had to face the reality that people who were often openly racist could also be incredibly kind to me—that racist people could be judged in most contexts as nice people. Yet some of their comments about Africans were so gross and incredible that I copied them down verbatim in a notebook, which I have reproduced in this book.

When I was in Africa in 1956–57, it would not be long before political winds of change would begin to sweep over this continent as the colonies I visited became independent nations. Unlike these countries, South Africa, although a member of the British Commonwealth, was completely self-governing at the time and in the process of implementing the horrendous practice of apartheid, meaning separateness. Under apartheid, which lasted until 1994, the government restricted non-Europeans to grossly inadequate schooling, inferior jobs, and substandard living space in slum areas, crowded reserves, locations, or "homelands."

Conversely, apartheid and colonialism had benefits for me, personally, which made my trip possible in ways it would not be today. Because whites were completely in control of the countries I visited, I was able as a lone white woman to travel everywhere in safety. Had I visited Africa five or ten years later, when, at least in the colonies, Africans were rising up in revolt against white domination, this would have been highly dangerous or impossible. Indeed, it says much for the Africans whom I met that even in the mid-1950s, when they routinely were treated badly by Europeans, they still were kind and friendly to a visiting Canadian woman such as myself.

Because I was white, I had an immediate entrée into the dominant class everywhere I went. I was always allotted good seats in a theatre and offered the best travel arrangements. Because I was a visitor to Africa, acquaintances were usually anxious to make me feel at home and like their country. The hospitality of the whites was amazing, and reflected in part the free time they had to perfect such kindness, given that the manual work involved in preparing meals, washing dishes, cleaning, and laundry was all done unobtrusively by African servants. The time and attention that my short-term hosts, Professors Jakes and Griff Ewer at Rhodes University in South Africa, gave to their graduate students and myself were staggering (and magical for all of us).

Although I thought of myself as a zoologist during my sojourn in Africa, I doubt if most other people did. I came from a population of girls who were taught that unmarried sex was sinful and kissing wrong unless you were engaged or married. Even if I had had access to birth control, I never would have considered sleeping with any man, no matter how beguiling. Probably few young women would be so naïve and puritanical nowadays.

That didn't mean that I refused to flirt with men or accept invitations for activities other than sex. Indeed, personal adventures which I considered too risqué for either letters to my mother or my journal (although they were relatively tame even at the time) and which I remember vividly even today, I include to give a richer personal flavour of what it was like to be a naïve young woman working and travelling alone in 1950s Africa. This memoir is similar in format to my book *Camel Quest: Summer Research on the Saharan Camel* (1978), which describes my adventures many years later with a fellow zoologist and many camels in Mauritania.

In this book I have kept the honorific names people used at the time, which reflect the extreme hierarchy between white and native: I and everyone else called native Africans by their first names (did they even have last names?) and they called me, if they called me anything at all, "Miss Anne." Every single white South African I encountered called native men "boys," so I have sometimes used this incorrect word in the text, just as I usually refer to young women such as myself as "girls." With the exception of Jakes and Griff Ewer, I called older white people "Mr.," "Dr.," "Miss," or "Mrs." and they called me "Anne."

# ACKNOWLEDGEMENTS

I am pleased to acknowledge those Canadian friends and relatives who, in the early 1950s, supported me as a young woman in my dream to study giraffe in Africa, especially my mother, my brother Donald and his wife, Wendy, Bob Bateman, Alan Cairns, Bristol Foster, Rosemary Hynes (Rowe), Judy Stenger (Weeden), and Mary Williamson.

Thanks also to those who helped me with this recounting of the year, especially fellow-author Sheelagh Conway, who thought of a way to recast the contents to make them more interesting. Other people have kindly read parts of the text and/or made useful comments: Yadhav Balakrishnan, Mark Behr, Alan Cairns, John Cairns, Wendy Campbell, Charles Gaulke, Barry Lively, Bev Sawyer, Elaine Sim, Mark Telfer, and anonymous reviewers connected with Wilfrid Laurier University Press.

I am most grateful to the people at Wilfrid Laurier University Press who made this book a reality—Brian Henderson, Carroll Klein, Jacqueline Larson, Pam Woodland, Catharine Bonas-Taylor, and Leslie Macredie—and to Kristen Pederson Chew who copy edited the manuscript.

And I shall always be grateful to all the diverse people, most of whom are mentioned in the text, who made my visit to Africa so magical. Thank you all.

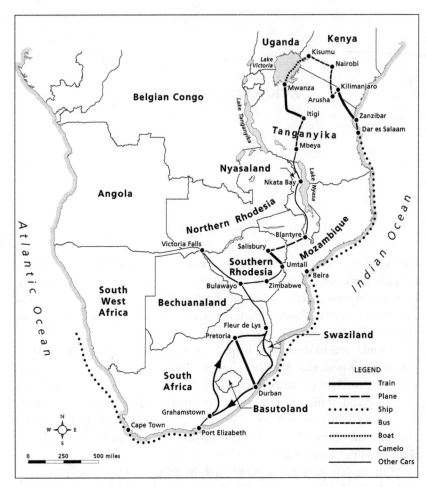

Travels in Africa.

# INTRODUCTION

It was the mid-1950s in Toronto, a world quite different than the one we know today. Toronto prided itself on being a city of churches, whose bells rang out musically on Sundays to lure us to worship. There wasn't much else to do on the Sabbath because there were no movies, no malls, and no stores open, although this didn't matter much to my friends and me who made our own fun. We lived in brick houses and were from cosy two-parent families, the mothers staying home to clean, cook, and shop while the fathers departed each day to their middle-class jobs.

My own family was a little different from the others because my father was a professor and Dean of Graduate Studies at the University of Toronto, the flagship university of Canada. My mother was not just a homemaker but a writer, too. Professors at the time weren't well-paid and we didn't own a car, but even so we lived in a large house on Dunvegan Road, just outside Forest Hill Village, for which my parents had paid $9,000 in 1942. Before the war, during the Depression, we even had a maid; I still own the little silver bell Mother had tinkled in the dining room when we had finished the first course of dinner and wanted the maid to bring dessert from the kitchen where she ate. All of these women had dropped their aprons and were gone like a flash as soon as war work and better wages became available.

Neither of my parents was born to such luxury. My father, Harold Adams Innis, was born in 1894 and came from a poor dairy farm in Oxford County, Ontario, the only child of four in the family to go beyond high school. On some days he went hungry while studying at McMaster University because he didn't have enough money to buy food. After graduating with a BA, he enrolled in the Canadian army to fight in the First World War at the age of 22.

He served for nearly a year in the trenches in France before being wounded in the leg at Vimy Ridge in 1917 and sent back to England to recuperate. Although he didn't like to talk about the war, he sometimes showed his children the small notebook with a hole in it through which a piece of shrapnel had passed to cause the wound; he walked with a cane for many years after. While teaching and working toward his PhD at the University of Chicago, he met and later married my mother, who was one of his students.

Harold Innis was an economic historian in the Department of Political Economy at the University of Toronto. He became famous for his original research, first writing on staple resource industries of Canada (such as *The Fur Trade in Canada: An Introduction to Canadian Economic History* and *The Cod Fisheries, the History of an International Economy*) and then on communication theory (*The Bias of Communication*). When he was only 58, he died in 1952 of cancer, but with the founding of Innis College in 1964, his name continues to live on.

Mary Quayle Innis was also the eldest and the only one of four children in her family to be engaged in academic pursuits. Her father worked for the Bell Telephone Company in a succession of cities and towns in the United States. Later, after my mother had left home, he and his wife spent years in Shanghai overseeing the installation of phones there. Mary Quayle attended the University of Chicago, focusing on courses in English, history, and economics, subjects which she continued to pursue after her marriage. While the mother of four small children (I was the youngest), she wrote a book published in 1936 entitled *An Economic History of Canada* because her husband wanted a text for one of the university classes he taught. This book garnered fine reviews and remained in print, used by thousands of students, for many years. For the rest of her life she continued to write—short stories, articles, a novel, and many books about Canadian history.

With my thoroughly Canadian background, why giraffe? Why Africa? I saw my first giraffe while on a visit to Chicago when I was two years old. Perhaps their height, especially from a small child's perspective, impressed me; perhaps it was the rush of movement when something startled them and they cantered in a flurry of necks and legs across their paddock. Whatever it was, the giraffe immediately became my favourite animal. I forgot the giant pandas with their trained nurse and the chimpanzees who drank tea sociably at a table set out on the lawn; all the other treasures of the Brookfield Zoo were nothing compared to the wonder of the giraffe.

As there were at that time no giraffe in Canada, I was unable to admire my idols in the flesh for many years. Instead, until I went to school, I contented myself with drawing pictures of my favourite animal; when I was able to read, I hunted for books on them. To my disgust, I found there were none other than those with quaint drawings for laychildren who had no sustained interest in the giraffe. I filled this gap by reading about smaller, inferior (although Canadian) animals such as the beaver, muskrat, and raccoon in children's books that I borrowed from the public library. I still have a filing card of notes on deer copied in a childish hand from information contained in *Bambi* by Felix Salten. None of these works could have compared with *Long Neck the Giraffe,* had there been such a book.

By a stroke of great fortune, the house on Dunvegan Road, which my parents had bought when I was nine, was only a block and a half from The Bishop Strachan School and had a girl next door, Mary Williamson, who would immediately become my best friend. When we were thirteen and ready for high school in 1946, our parents enrolled us in this private girls' school where we were taught to be modest, frugal, and always to act like ladies. I adored every minute of my five years there. I loved sports, I loved my studies, I loved my family and our dog, Tigger, and being outside to bicycle and read in the summers; there was no time in my life for politics or foreign affairs of any kind. But most of all I continued to love giraffe.

When I was seventeen, my mother and I made another pilgrimage to the Brookfield Zoo, our first destination, of course, the giraffe house. The giraffe were still magnificent. They were less tall than I remembered, nor did they move as quickly, but these were details. A lordly male leaned over the fence and stared down at me with huge brown eyes; I gazed with reverence back at him. My passion for giraffe was firmer than ever. We retired finally at the insistence of my mother, whose mania for animals did not equal my own. As we departed, we glanced with awe at the height of the giraffe's fence compared to that of their keeper, and at the size of the giraffe house itself. Both were colossal.

When I arrived at the University of Toronto in 1951, the giraffe was still my first love. Rather than study modern languages or history or fine arts as my friends did, I enrolled in the honours science program which would lead to zoology and, I hoped, the giraffe. During my first year I struggled with English, French, geology, physics, chemistry (eight hours a week), mathematics, and botany as well as zoology; in the zoology course our class concentrated on the amoeba, and progressed painstakingly via worms and frogs to the rabbit. Later, we would memorize the scientific names of countless clams, fish, and insects, and each of us would prepare a plant collection of a hundred specimens glued onto sheets of paper, but we learned nothing about giraffe.

During my first nine months at university, my father was slowly dying at home of cancer which made this period even more difficult. Every evening when I returned from university, walking up Dunvegan Road from St. Clair Avenue, I watched anxiously to see if his bedroom light was on. If the room was dark, I imagined that he would have died. If it was bright, I thought again of the terrible pain he was suffering and of how helpless I felt to do anything about it.

University life then was quite different than it is now. Like my girlfriends I always wore skirts, saddle shoes, and bobby socks to class—slacks (the ubiquitous jeans were still in the future) seemed to me then to indicate a lack of respect for the university, and indeed it was forbidden to wear them at dinner in the women's residences. We rode the newly opened Yonge Street subway to campus, thrilled at the time we saved compared to travelling everywhere by streetcar. For the first time in my life I went out with boys, usually in groups of classmates but often as a couple, too.

None of us students had weekday non-academic jobs to interfere with our university studies—tuition was low, about $150 a year, and fast-food restaurants, which now supply work for so many, unknown. On weekends I earned twenty-five cents an hour reshelving books at the Deer Park Public Library; one memorable week, without warning, my pay, with the implementation of a minimum wage law, leapt to the heady sum of forty-three cents an hour. During two summer holidays, donning a white lab coat, I filed research tissue slides and acted as a "gopher" at the Banting and Best Medical Research Institute on College Street for twenty dollars a week. The Institute was named for Frederick Banting and Charles Best, co-discoverers of insulin, who with others made the university world-famous with their discovery in 1922. Charles Best, in an expensive suit, occasionally came into our lab and preened about looking pleased with himself. The Institute was at the cutting edge of research; one Saturday I returned to the building to watch an early experiment diverting the blood of an anesthetized dog through a lung machine so that a surgeon could work on its (healthy) heart. Such beginnings would soon lead to successful open heart surgery.

Unlike today, the faces on campus then were virtually all white, including the many Jews in my college, University College. I was unaware that some Canadian universities had quotas for Jews, and that Jewish scholars were refused employment because of anti-Semitism. One of my friends, Bobby Kajoka, was of Japanese background, but I didn't realize then the conditions under which his family had suffered when Japanese-Canadians were interned during the war, their Vancouver-area homes and businesses expropriated without compensation. I don't remember any black people at all, even though I was sensitive to diversity and attended meetings of various exotic groups to broaden my mind. Once I arrived at the door of a Buddhist (or was it a Baha'i?) meeting to find two men wearing full beards in the room, something I had never seen before among my peers. This, to me, was *so* exotic, if not actually scary, that I decided to skip the meeting.

After graduating with a BA and the Gold Medal in biology, I longed to leave for Africa to fulfil my dream of studying giraffe there. However, although I wrote to game departments and officials in Kenya, Tanganyika, and Uganda to try to arrange this, none could help me. Nowadays a student would go to one of her professors and ask which universities would be most generous in financing graduate work in Africa. In the 1950s, however, there was no tradition of zoology students or professors doing field research overseas and, of course, no money or infrastructure for this. Indeed, I was apparently the first zoologist to go to Africa to carry out a long-term scientific study on the behaviour of a wild mammal. I would be followed by men studying kob antelope, baboons, gorillas, wildebeest, and, four years later, by Jane Goodall observing chimpanzees (see Appendix 2). After these would come thousands of zoologists, psychologists, and anthropologists to do further research on wild animals. However, back then if I did mention my ambition to study giraffe in Africa to faculty members or anyone else, they laughed heartily at its absurdity. Later, Professor David Fowle wrote me, "I recall the amusement with which your announcement of your intention [to go to Africa and study giraffe] was greeted by your friends."

Finally, not knowing what else to do, I spent another year at the university earning a master's degree in the genetics of mice; my friends had a wonderful time joking about research on mice as a prelude to studying giraffe. At this point my mother, with her four children almost grown, was preparing to support herself by becoming Dean of Women at University College. She moved from the apartment on Bedford Road we had shared, after she sold the Dunvegan house, into two rooms in the Women's Union on the University of Toronto campus as part of her new position, so I became a boarder in Falconer Hall, one of the female graduate student residences.

Eventually I made contact with a professor, Dr. Jakes Ewer, in South Africa and decided to go there as soon as possible because he had heard about a farmer who had giraffe on his cattle ranch and might be able to help. He had written about me to the farmer, Mr. Matthew, but had given my name only as A.C. Innis; Mr. Matthew, assuming that the student was a man, had said he would be willing to have him stay on his farm. I had saved money from working summers not only at the Banting and Best Institute but also at the Royal Ontario Museum, cataloguing animal specimens, feeding mice, cleaning bones, and measuring fox skulls—nineteen measurements on each of one thousand skulls. (My male classmates had more exciting and lucrative summer jobs doing field research, jobs denied me and my girlfriends because we were female.) During my senior years I had also earned money demonstrating for five different laboratory courses in biology. With a scholarship and a gift of money from my mother, I would be able to reach Africa and probably return as well.

In May 1956 I set sail from Montreal for London—at that time plane travel was prohibitively expensive and uncommon. From there, I wrote to Mr. Matthew to say that I was most anxious as a trained zoologist to observe the giraffe on his ranch, but that I was not a man as he had assumed, but a woman. Then, before he could answer and possibly refuse me because of my sex, I sailed for South Africa, asking him to reply to me care of Dr. Ewer at Rhodes University.

# 1

## *Setting Off*

When I see him in the lounge with his friend, I settle down behind them to read my book. If the two of them are strolling around the deck, I stroll nonchalantly after them, hoping they won't notice me. I only catch snatches of their conversations, always innocuous.

"I like having tea on deck."

"The students behave better in Rhodesia than in England."

"Look at that sunset!"

They probably don't notice me because I'm white, like all the other passengers except them. He is tall, handsome, and black; his friend is shorter, not very good looking, with a lighter skin colour. It's good that they have each other to talk to because everyone else gives them a wide berth. We are all sailing to South Africa on the *Arundel Castle,* a two-week trip from London. It is July 1956, a time when blacks and whites inhabit separate worlds.

I've read enough issues of *National Geographic* to be entranced with the romance of black Africa where my beloved giraffe live. As a Canadian I've only seen a few black people in my entire life and never spoken with one, so no wonder I'm desperate to talk to this exotic man so close at hand. I don't have the nerve to close in on two of them at once, though. I bide my time.

At last, several days out from England, I see him standing by himself at the rail looking out to sea, dressed as usual in black pants and a long-sleeved white shirt and tie. Behind him white people sprawl on deck chairs in the sun in swimsuits, shorts, halter tops, and tee-shirts, working on their tans.

I saunter over to the rail on his right, not too close to scare him but close enough to chat.

"Isn't it a gorgeous day?" I say, gazing also at the blue water. Sometimes I'll see porpoise, flying fish, or albatross from the railing, but today there is no sign of life.

He starts. "Yes," he agrees.

"Do you live in Africa?" I ask, turning my head to look at him. He does seem nervous. He shifts his feet.

"Yes, I'm from Southern Rhodesia."

"Have you been away long?"

"I've been teaching in England for a year," he replies.

"Are you glad to be leaving?"

"Not really," he says. "The English are good in that most don't make a big thing about skin colour. Many English people were friendly *because* I'm an African, which never happens in Africa."

I'm beginning to sound like an inquisitor, so I say, "I'm from Canada. My name's Anne." I offer my hand which, after a second's pause, he shakes.

"I'm Josiah Chinamano."

It's delicate work, but I manage not to frighten him away and to extract a bit of his history. He lives in Salisbury with his family, where he is superintendent of thirty mission schools in the area. The past year he has been lecturing on Africa at Selwyn Oaks College in Birmingham. He is religious and deeply interested in the racial climate and politics of Africa.

Josiah is warming up to me. "Why are you going to South Africa?" he asks.

"To study giraffe," I say. "I'm going to write a book about them because there isn't one. There's a ranch in the eastern Transvaal where I may be able to watch them."

Josiah wrinkles his nose. "Aren't there better places further north to study giraffe?" he asks. "Kenya? Tanganyika? Uganda?" He seems to have little use for South Africa, which I'll understand when I get there.

I'm delighted with his interest. "I wrote to everyone I could think of for help, but no luck," I reply. "Mr. Cowie, director of the Royal National Parks of Kenya, wrote that Kenya was out because there were still a few Mau Mau revolutionaries murdering people in the countryside. He's the guy who years ago wrote a letter to a Nairobi newspaper signed 'Old Settler,' arguing that for the sake of progress in Kenya they should machine gun all the game animals and poison all the predators; as he intended, the huge backlash to this idea jump-started the push for conservation. Anyway, Cowie forwarded my letter to Dr. Leakey at the museum in Nairobi. Dr. Leakey was sorry but he knew of no jobs I could do to help finance my field work, and no real base I could use." (Four years later, Leakey would set up Jane Goodall in Tanganyika so that she could begin her world-renowned study of the behaviour of chimpanzees.)

"I've heard of Dr. Leakey," Josiah says.

"The Director of Education in Kenya had no good suggestions except to contact Mr. Cowie," I continue. "When I wrote Mr. Cowie again, he wondered if I could study giraffe in Nairobi National Park although there's no research base there. The Survey of Tropical Africa thought Dr. Leakey or the East African Game Departments might be able to help me, which they couldn't. A teacher at the London School of Economics recommended a professor at Makerere College in Uganda as a possible sponsor, but this didn't pan out either."

"How did you get the Transvaal connection?"

"This all happened last summer," I say. "By the time fall came, I still had no hope of getting to Africa so I spent last year working on my master's degree at the University of Toronto."

"You have a master's degree?" Josiah asked, startled.

"Yes. I wrote a thesis on the growth of newborn mice, showing that different genetic strains gain weight at different rates when a set diet is fed to their mothers. It was the only research project available that didn't involve hurting animals. Not much use for giraffe research, though," I laugh.

"Late last fall, a friend told me about a giraffe named Shorty in the Hluhluwe Game Reserve in Natal," I continue. "He was born in the Transvaal, abandoned by his mother when a few days old, and hand-raised by a farmer who bedded him in a stable and allowed him to wander near the farm house. When Shorty was six months old—over eight feet tall!—Transvaal game authorities expropriated him because giraffe are protected game and can't be owned by a private citizen. They sent him to Natal in exchange for other animals, hoping that he would increase tourism at the Hluhluwe Reserve even though giraffe never occurred there naturally.

"He used to wander among the huts in the rest camp, eating hay and titbits from the guests and posing for photographs, but one day he kicked a girl bending over to take a picture of a flowerbed. From then on he was barred from the rest area and spent his days moping about outside the camp gate. I'm not boring you, am I?"

"No, go on. What happened then?" Josiah asks with some interest.

"A zoologist from Rhodes University in South Africa called Griff Ewer was visiting Hluhluwe. When she commented on Shorty to the camp warden, he told her the animal's sad story. Shorty's original owner had a ranch next to that of an Alexander Matthew, whose property, Fleur de Lys, had about ninety-five wild giraffe. Mr. Matthew had indicated that he, too, would send giraffe to Natal if that would help increase the range of the species; he was interested in wildlife and realized the importance of carrying out field research.

"About this time a fellow graduate student of mine from the University of Toronto [and later a professor there], Rufus Churcher, gave me a letter of introduction to his former teacher, Dr. Jakes Ewer, so I wrote to him at Rhodes University. He contacted the National Parks Board of South Africa for me, but this Board didn't want giraffe studied. Then Griff returned to Rhodes and told her husband, Jakes, about Mr. Matthew. Jakes Ewer wrote to him at once, explaining about the Canadian student who wanted to study giraffe. To be on the safe side, he didn't mention that I was a girl."

Josiah makes a face at this, but I continue anyway. "Mr. Matthew was wonderful. He wrote Dr. Ewer that he'd be glad to assist A.C. Innis, offering him lodging for a small sum or part-time work if he wanted to earn room and board in a men's bunkhouse. He said he had riding horses and a Land Rover that Innis could use to watch the giraffe."

"What did Mr. Matthew say when he realized you were a girl?" Josiah asks.

"He doesn't know yet. I wrote him just before I left London and gave Rhodes University as my next address."

"Was Dr. Ewer all right with that?"

"He wrote not to worry about such details. He said I should come along to South Africa and sort the matter out later. I hope it works out. I'm not game for bunking down with a bunch of stockhands."

Josiah can't help laughing, although he glances around to see if anyone is watching him having such a cheerful conversation with a white woman.

As this point his friend, Mr. Chetty, joins us. As we three chat I realize that Mr. Chetty, a so-called "Cape Coloured" because some of his ancestors in Cape Town were white, is not the least concerned about the colour problems of Africa. His goal is to immigrate to Britain. It turns out that the two men do have a white friend on board ship, a Rev. Illsley, who runs a mission in Southern Rhodesia and has just published a novel called *Wagon of Fire*. When I later meet the minister he is most friendly, insisting that I come to visit him at his mission where his wife can put up as many as eighteen guests at once.

I think of Josiah as a friend, but he makes a point of talking to me only on deck and only if there are few other people around.

"Are you coming to the dance tonight?" I ask him once, to be friendly; every evening the small band plays on deck under the moonlight.

"No," he says. "I'll be reading in my cabin."

Because we are both interested in books, he borrows from me *Naught for Your Comfort* by Father Huddleston, an Anglican priest who is dead set against apartheid in South Africa. In return, I borrow from him *Wagon of Fire*—a grim thriller bulging with adjectives about natives versus Christianity versus whites, with Christianity apparently winning. Before sailing, I had visited the Charing Cross area of London to browse through second-hand book shops, hunting for animal books that included information on giraffe. There were none. Instead, I bought eight books on Africa to read on the ship. Next to the giraffe, I'm anxious to learn as much as I can about the continent that I've dreamed about for so many years, and its human inhabitants.

When I'm reading one of my new books on deck, I'm often served free lemonade by a deck steward from Nova Scotia, the only other Canadian on the ship. This way he'll have an excuse to stop and chat. Reading a book seems to be a signal for anyone to come and interrupt, as if this activity is a last resort and that any human contact at all is preferable. The steward introduces me to the Master-at-Arms, who has two sons working in Canada, so the three of us can discuss our common interest.

"I love to hear the way you talk," the steward usually says when we meet.

"I love not to have to keep repeating myself," I respond; most of the English and South African people I talk to have trouble understanding my Canadian accent. I'm also pleased with the free drinks as I don't have more than just enough money to last the year. Once, when I'm talking to the steward, the ship's photographer insists he'll take my picture on spec. I turn my head to prevent this, but he takes it anyway and, alas, when I see it, even without my face, I'm too conceited not to buy it.

My acquaintances on board are, with the exception of Josiah and Mr. Chetty, a cross-section of white people going to Africa. One man, an engineer, falls into step beside me one dark evening as I'm walking round and round the main deck for exercise.

"Have you seen the luminous plankton?" he asks when he finds out I'm a zoologist.

"What's that?" I've never heard of such a thing, but it sounds biological. I'm immediately interested, and he is delighted at my ready response.

"Come on, I'll show you," he says eagerly.

We walk past a couple giggling in a deck chair and to the very back of the ship by way of the crew's quarters where, by leaning over the rail, we can see pinpoints of light flickering in the surge of dark water below churned up by the ship's propellers. I find this vision entrancing, at least until the engineer puts his arm around my waist and pulls me close to him.

"What are you doing!" I say, startled.

"The plankton isn't *that* interesting," he insists, pulling me around to face him so he can kiss me.

"Leave me alone," I snap, pushing him away. I immediately retreat the way we came, past the deck chair where the giggling has been replaced by frenzied creakings, and into the lighted lounge. Why would anyone want to kiss a person they'd known for only five minutes? Was this shipboard culture? I'd been taught by the chaplain at my all-girls high school that one should only kiss a boy one is engaged to. This had seemed strange: Wouldn't kissing lead to an engagement rather than the other way around? A commonly accepted rule of thumb in university at that time was not to kiss a boy until the third date. This seemed reasonable then, although incredible now.

During the trip I become friendly with three travelling groups, the most empathetic being seven male botanists employed by the British colonial service to teach agricultural methods to the Africans of Nyasaland (now Malawi). (Why would farmers who have spent a lifetime growing crops in their native land need such help, I think now.) To accustom the botanists to the tropics, the British government had sent them to Trinidad for a year to study agricultural conditions there. Now they are heading for Nyasaland to share their new-found wisdom with the local farmers. They spend each day on ship struggling to master the intricacies of Chinyanja, the language spoken in their new country. They're much more serious than the other British colonial servants going to the Federation of the Rhodesias and Nyasaland for a three-year tour of duty, or South Africans returning home from their holidays.

The second cohesive group is an enthusiastic rugby team of twenty Welsh schoolboys scheduled to play a series of matches with South African teams. In the evening I often play a game of Categories with some of them. We each have a piece of paper on which we write across the top categories such as makes of cars, kitchen appliances, and kinds of dogs. Down the side we write five letters, then try to fill in each of the squares of categories matched to each of the letters—a game now marketed as Scattergories. My opponents usually nominate a category for Welsh rugby players for which I prove adept, despite complete ignorance of the subject.

"White for W," I state confidently as we go around with our answers.

"I don't remember a White," one lad will say after thinking for a moment.

"No, Rod White," another will break in suddenly. "Remember he played about ten years ago? For the Cardiff team?" The others agree that White is a good choice.

"J for Jones," I say next.

"Everyone has him," they chorus. "Brian Jones, he was great."

"Didn't he play back?" I ask tentatively.

"No, no, no, he was a forward," they insist, thinking how dense I must be not to know that.

Before our games they spend the afternoon practising rugby moves on deck. When they are finished they gather, hot and sweaty, to sing Welsh hymns and songs in parts as they cool off.

Compared to these fine lads, the men in the third group are repulsive—seven British policemen hired to work in Southern Rhodesia, the most notable of them famous for having drunk twenty-nine bottles of beer in one evening. By agreeing to serve in the colony, some of the younger men avoid their two-year term of National Service. Early in the trip, before I knew what he was like, I had shown Paddy, from Ireland, my stateroom because he had asked to see it. He had pushed me down onto the lower berth but let me go when my roommate, Shirley, suddenly came into the cabin. Had he been about to try and rape me? I don't know, but he was interrupted so promptly that I didn't dwell on his behaviour, except to never again be alone with him. I couldn't avoid him entirely because the common lounge space was too small.

From their conversation I pity any native law breakers these policemen will deal with. Although they have never been to Africa, they act as if they know all about Africans.

"We've seen you talking to that coon," both Paddy from Ireland and Derek from England say to me on different occasions. "What do you talk about?"

"About Africa. What else?" I reply lightly.

"Be careful you don't encourage him," Derek says. "If I see you dancing with him, I'll kill him."

"Don't be ridiculous."

"He probably thinks he's smart because he's a teacher," Paddy says, "but you can't teach a coon anything. They're like four-year-olds."

"Maybe seven-year-olds?" Derek laughs.

"That's silly," I say.

"*I'm* not putting up with any nonsense," Paddy states, making a motion as if cracking a whip. "They'll know who's boss when I'm around." He laughs, sits down beside me on the sofa and puts his arm around my neck, pulling my head forward to give me a kiss. I struggle free.

"Don't," I snap. Paddy's friends laugh, to Paddy's annoyance.

"Anne's stupid too," he says. "She reads books but she doesn't know anything. She lifts up the tail of a horse to see what's underneath." I stand up and stalk away from the group; there's silence behind me, then a burst of laughter.

Usually, when I'm relaxing by myself on a deck chair (no sunblock then, alas), I worry about Ian. Have I lost him for sure because of my dream of studying giraffe?

I met Ian in June of 1954 at the Timothy Eaton Memorial Church Tennis Club, a handsome guy of twenty-six who had a PhD in physics and was doing research at the University of Toronto until he left in the fall for a one-year fellowship at Oxford University. I was thrilled when he asked me out the next evening to see the movie *The Student Prince,* then kissed me on the doorstep of the apartment where I lived with my mother. I'd never been kissed on a first date before, but I didn't mind at all.

From then on we were together every evening that summer, playing tennis, seeing a movie (he always paid) where we held hands, or hanging out with his friends. He didn't have a car so I often drove us in the family Austin which my father had bought two years before he died (though my parents never learned to drive it).

"You shouldn't spoil boyfriends by driving them around," Ian admonished me, but nevertheless he liked not having to depend on streetcars.

When I found that he'd briefly dated four of my acquaintances, I realized he must prefer me to them. One of these, Diana, wrinkled up her nose when she learned I was going out with Ian.

"He's awfully fast," she sniffed, probably referring to a first-date kiss.

"Oh, he's all right," I said in an offhand way. How exciting and unlikely to be thought a vamp! My brother routinely called me "little brown hen," so I was amazed that Ian had chosen me over my friends.

When Ian left for England he promised to write me, which he did now and then. I wrote back gladly, always ending each letter with "Old Flame News" about the four women he had dated; this had the double advantage of making him seem wonderfully popular and me superior to these rivals.

The following spring, after my BA graduation, I spent the summer with my friend Anne Dalton, travelling through Europe by train and hitchhiking around Scotland and England. In Oxford, Anne and I had dinner with Ian at an Indian restaurant. After we walked him back to his college Anne returned to our hostel, leaving Ian and me alone by the gate.

"See you much later," she called back to me, teasing.

I moved closer to Ian, hoping for a hug, but he just stood there. "I'm seeing an English girl," he said. "She's studying zoology, like you."

"Oh," I said, stepping back again.

Minutes later I joined Anne at the hostel.

"That didn't take long," she remarked.

"He's got another girl," I admitted glumly.

However, our brief meeting acted as a catalyst for letter-writing because he wrote me a few letters in August and September before another lapse. Many of my friends had steady boyfriends, so I liked to think that I had one too, even though he didn't write often.

"Never mind," I shared my feeling of being neglected aloud with my new friend at graduate school, Dorothy Armstrong (a future Canadian ambassador), who lived in the residence room next to mine at Falconer Hall. "He'll be busy moving back to Canada. He'll show up soon."

There was no sign of him in October 1955, but he finally arrived in November. He called at the residence late one afternoon looking for me and was directed by my friends to the third floor of the Zoology Building. I was in the daily process of weighing hundreds of baby mice for my thesis research.

"I found you," he announced, pulling up a stool beside mine, but not smiling.

"Hi!" I exclaimed, relieved and thrilled to see him at last.

"I wanted to tell you I'm engaged to a Dutch girl. We're getting married next summer."

I looked at him aghast, unable to speak.

"I've got a job at the National Research Council in Ottawa. My mother and I are driving there from Winnipeg now with all my stuff, so I can't stay."

When he'd gone, I continued weighing each tiny body, keeping my head down so that nobody would notice me sobbing. I wiped away tears that fell on the babies so their weight wouldn't be affected. I was devastated.

In the evening I slunk back to the residence in a state of shock. My friends, sitting around on the floor in the common room, burst into song when they saw me come in—Wedding Bells Are Ringing for That Old Gang of Mine. I gave a hollow smile and mirthless laugh before retreating to my room. The singing outside died down uncertainly. After a minute, Dorothy knocked on my door and came into the room.

"What happened?" she asked, her face twisted in sympathy.

"He's engaged to a Dutch girl. They'll be married next summer."

Fast forward to an explosion several months later in which several people at the National Research Council in Ottawa were killed. When I heard this news on the radio I was so upset that I had to sit down to recover. Ian might be dead! I must still care for him, I thought, so I sent him a note saying that we should still be friends even if not romantically involved. (What a cliche!) He wrote a letter back: "Although I won't admit that I was wrong about members of the opposite sex not being good friends—I will say that there are exceptions—and I pronounce that you and I are exceptional—in more ways than one of course."

Ian visited me over Easter of 1956 when he was in Toronto, and we spent a wonderful two days together during the Victoria Day weekend in May when he was best man at a friend's wedding; I was invited to the ceremony at Ian's instigation. Before he returned to Ottawa, we took a streetcar to the Riverdale Zoo and sat on adjacent logs near the deer enclosure.

"It's really you I love," he told me, holding my hand while a doe stared at us through the fence. "I don't want to marry anyone but you."

I was stunned. What should I say? Ian was proposing marriage, and I was getting ready to sail to England in less than two weeks.

"I don't know what to say," I blurted out. How could I get married if I wanted to go to Africa?

"Come back with me to Ottawa so we can discuss it," he said.

I took the train to Ottawa the next weekend to try and figure out what to do.

"Why don't we spend a few months getting to really know each other?" Ian said. "We've only had one short summer together nearly two years ago."

"But I'm booked next week to go to England on the way to Africa to study the giraffe. I've always wanted to do that."

"Maybe we could go together to Africa later on," he suggested. But what would a physicist do there?

"Would you wait a year for me?" I asked.

There was a long pause. "Of course," he said finally.

But would he? He was everything I wanted in a man—fun, well-educated, handsome, good at tennis, generous. I didn't really want to get married, but would I later regret letting him get away if I went to the giraffe? (Women in those days talked about catching a man, and not letting him escape.) I didn't ask about the Dutch girl. Were they still engaged? Maybe he would go ahead and marry her anyway?

I asked my mother what to do. As a Dean of Women used to giving counsel to troubled young women, she suggested I go to England on May 31st with my friend Rosemary Hynes, as planned, and make a final decision there—return to Canada to spend time with Ian or carry on to study the giraffe in Africa. I was so hugely conflicted that on the ocean crossing I was deathly seasick for the only time in my life.

My confusion wasn't helped when I picked up my mail at Canada House in London to find a letter from Ian waiting for me. He wrote that he'd asked his Dutch girlfriend to visit him for a month or so in Ottawa to sort things out. My heart sank. Tension about my future began to make my left eyelid involuntarily flutter sporadically. Later, a second letter arrived from Ian saying only that his Dutch friend would not be coming to Canada.

What to do? Try for Ian or opt for the giraffe? The giraffe or Ian, the good catch? With great luck I could spend the year following giraffe and then have Ian. With less luck I could have the giraffe but *not* Ian. Would I regret this for the rest of my life? Or if I chose Ian rather than the giraffe, would I resent him later on? I toured around Britain for over a month with friends, while mulling over these thoughts. My whole future depended on the choice I made.

Unlike today, when ocean liners are usually cruise ships carrying hundreds of passengers on short-term holiday, in the 1950s liners were the main mode of transport across oceans. The Union Castle Steamship Line ran a ship from London to South Africa once a week, leaving on Thursdays. I asked at the office of the company for an early cancellation on a ship sailing to Port Elizabeth in South Africa, giving the name of the hostel where I could be reached if a berth became available. If there was no vacancy for a month, I resolved I'd return home. If there was a cancellation, I'd go to the giraffe. I would let fate decide.

The next day the giraffe won. A Union Castle official phoned to say that I could leave in two days, on July 19th, 1956, for South Africa if I brought a certified cheque for £60 (about $300 CDN) to his office. I rushed my cheque to a bank and then downtown, packed my suitcase and backpack, and waved goodbye to Rosemary who would return alone to Canada.

Stretching in the deck chair, I notice my roommate Shirley and call to her to join me. She's a travel agent who has been to Britain on a visit; she sleeps in the lower bunk in our tiny stateroom with its small porthole. Before our sumptuous dinners we often have a drink, but first I wait for her to get dressed for the evening, always in a new ensemble I haven't yet seen. She rushes down the corridor to the women's washroom to shower, then presses her skirt in the ironing room on the way back to our cabin. Then she chooses matching earrings from her jewellery case and puts on nail polish, mascara, eyebrow pencil, powder, and lipstick. She is so completely man-oriented that when I mention that I long for mail, she assumes I mean "male" and want a date. Of course, I *am* thinking about a letter from Ian. I've sent him the ship's address and schedule of ports of call from London, hoping there will be a letter from him in Cape Town.

When she is ready, her new boyfriend comes by to pick her up. Andy, who lives in Durban, has no more use for Africans than do the policemen.

"*Kaffirs* are dirty, lazy, and stupid," he proclaims.

"That's a ridiculous stereotype," I say, noting Josiah out of the corner of my eye going to first-sitting in the dining room, a book on Africa I lent him tucked under his arm.

"You'll soon lose your idealism when you see what they're really like," he warns. I manage to restrain my anger with difficulty.

Andy then insists that I borrow a pamphlet explaining how Africans can excel at university even though they're stupid. According to the pamphlet they have a special photographic memory which enables them to answer questions on tests and exams even though they are totally unable to reason or understand what they have learned. Such propaganda must be important in brainwashing people such as Andy to believe that oppressive laws for natives are justified.

While Andy and I argue, Shirley sits beside Andy, admiring her nails and saying nothing. If Shirley has a nap or goes to bed early, Andy takes up with Valerie who is happy to sleep with him according to Andy's roommate, Alec. When Alec wakes up there are sometimes two bodies in the other bunk.

"Andy wants to marry me," Shirley tells me.

"But you've just met him," I object.

"He really loves me. I know, because he ignores Valerie even though Valerie chases after him." Poor Shirley.

We arrive early in the morning at Cape Town, surely one of the most beautiful cities in the world. The buildings are white, with red roofs glistening in the early sunlight. Behind them rises Table Mountain to a height of 3,500 feet; at intervals during the day, white fluffy clouds known as the Table Cloth pause at this summit before drifting on.

All but forty of the passengers are leaving the ship here, most of them to take the train north to Johannesburg, Pretoria, and the Federation of the Rhodesias and Nyasaland. I loiter near the gangplank for a while, saying goodbye and good luck to the various people I've chatted to during the trip. They are all extremely kind: the minister urges me again to come and stay with him and his wife at their mission for Christmas; I'm given the names of three people in Grahamstown, whom I must look up (but don't); and two couples from my table give me their addresses in Johannesburg and Pretoria in case I visit those cities. (A year later I do contact one of these couples, but by then they have forgotten who I am.) In the background the Welsh boys sing sad songs but I remain in the best of spirits. At last I'm in Africa, the land of my dreams! If I had known then what we know now, that Africa is not only an exotic land but the home of our most distant human ancestors, would I have been even more thrilled? I doubt it. That concept would have been too strange, too foreign to be grasped.

Later, from two decks above, I watch Josiah, dressed in a suit, walk sedately down the gangplank with a small suitcase in his hand. We had exchanged addresses the night before and agreed to write to each other. Behind him come the policemen in windbreakers, punching each other playfully, shouting comments back and forth. I picture Josiah and the policemen unhappily confined together for the next several days on the train to Southern Rhodesia, and realize that I've learned a lot about Africa already. When Josiah refused to go to the evening shipboard dances, I thought it was from Christian scruples or an aversion to frivolity. Now that I'm aware of the vast sea of racism around me, I realize that, unlike the rest of us, he has to constantly monitor his behaviour just to be physically safe. How strange that English people in England made him feel welcome, while English (and Afrikaner) people in Africa, his homeland, treat him with contempt. Are the dregs of Britain especially attracted to Africa because the standard of living is far higher for them here than at home?

Five years later, Josiah will visit me in Waterloo, Ontario, and meet my new husband, who teaches at the University of Waterloo. I'm working on my definitive book about the giraffe, which will not be published for another fifteen years. When I go for a walk with him around Waterloo Park, people stare to see us together. We take him on Sunday to St Andrew's Presbyterian Church, along with my mother, where the congregation is thrilled to have a black man in their midst. Two male elders approach us to introduce themselves after the service. They invite Josiah and my husband to a church meeting the next night about missionaries and their work. Even though Josiah is *my* friend, I'm not invited and have to stay home instead. To my annoyance I realize that, in this church, being black trumps being a woman.

Later, Josiah becomes Joshua Nkomo's right-hand man in Southern Rhodesia's fight for freedom from British rule. Soon after Zimbabwe is founded in 1980, Nkomo is demoted from his cabinet position by Prime Minister Robert Mugabe, who becomes in effect the repressive leader of a one-party state, which he remains to this day. Nkomo's demotion also ends Josiah's political career.

In one of his letters to me, after his visit, Josiah asks about the possibility of sending his daughter, one of six children, to Waterloo to train as a nurse. This doesn't work out. We lose touch during the upheaval in his country, but I worry about him after I hear he is imprisoned by the white government in 1964 and featured by Amnesty International as a Political Prisoner of the Month who needs help. I send him money, as do some of my friends, and write letters to him and on his behalf, but it seems too little. I don't even know if he receives all my letters. I've heard recently that he has died, although his memory lives on in Harare, which has a main street named after him. His widow seems currently to be a Member of Parliament in the country; I've written to her but received no answer.

After immigration formalities on the ship are completed, I rush to the Purser's Office to see if there is mail for me. Nothing. No letter from Ian and none from my mother. Damn. Even so, when I troop down the gangplank to spend the day in Cape Town before sailing with the *Arundel Castle* to Port Elizabeth, I'm thrilled to at last set foot on African soil. I've been waiting for this moment for years. I wander happily through the city, listening to people speaking both English and Afrikaans, a bilingual mixture rare in Canada outside Montreal. (Afrikaans evolved from the languages of early settlers in South Africa, mostly Dutch, but also some French and German). When I board a trolley bus heading toward the mountain, the seats aren't segregated for blacks and whites. After leaving the shopping district we head into residential areas. At first we pass neat bungalows with small gardens where white people live; farther back are much smaller dwellings, attached to each other and without gardens. This is the area of non-whites, either Cape Coloureds (who have both white and black forebears, often from hundreds of years earlier), Indians, or Africans. At the end of the bus line I saunter farther uphill, enchanted to be by myself for a change. There is no noise from the city below, only bird song and the swish of wind through the pine trees.

In the afternoon I wander around the cultural centre of town, the lovely pink and cream provincial government buildings, the botanical gardens, the art gallery, and the museum of South Africa. Dr. Ewer has kindly written to Dr. Crompton, the director of the museum, who takes time from his work to chat with me for a few minutes. We stroll through various exhibits explained by legends in both Afrikaans and English. He hopes to give away a stuffed giraffe standing forlornly in a corridor; I'm tempted at the thought of having my very own giraffe, but even I realize it is too big and too tattered to be carted home to Canada.

In the nearby castle a guide offers to show me and two Afrikaner men the dungeons. After the men speak to the guide in Afrikaans, the guide turns to me.

"The men want to know where you are from," he says. I'm astonished. How can anyone live in South Africa, part of the British Empire, the ubiquitous red on world maps, and yet not know English? Andy tells me later that some Afrikaners just pretend not to know English to be ornery.

Shirley, Andy, and I are among the few people left on board. We rattle around for an evening before more passengers arrive to sail east along the coast. Among them, to Shirley's dismay, is her fiance, David, who has come to join her on the ship for the next few days until they reach East London. During my last evening on the ship I fall asleep to the nervous conversation of Shirley and David as they hash over their future while squashed into the bunk below me.

"But you're engaged to me! Andy is only a fling. You've only known him two weeks!"

"But he says he loves me."

"I love you too, and have for years."

"I don't know what to do."

After we dock at Port Elizabeth, I'm met mid-morning by Dr. Ewer. He has written that I'll recognize him by the copy of the magazine *Nature* he carries under his arm, but I don't need to know this detail. Unlike anyone else gathered on the pier he looks exactly like a professor—in his forties, brown-grey hair, rumpled pants, tweedy jacket, with an amused yet cynical expression on his face. He is pacing back and forth as if in thought, giving a skip every now and then.

"Welcome to Africa!" he says and introduces himself, delighted to find that I have brought only one suitcase and a knapsack to last me a year. The other girl he's driving back to Grahamstown, a student in education who has spent the past two weeks doing a practicum at a reform school in Cape Town, has four suitcases. She gives a one-shilling tip to the African worker who hauls her luggage to the car and stuffs it into the boot.

# 2

## *Adapting to Africa*

The eighty-mile drive to Grahamstown from Port Elizabeth is along a two-lane paved highway winding north-eastward over low hills through scrubby vegetation. Few signs of white people sully for me the essential Africa beyond several fields of pineapples and four farmhouses, each with two or three round thatched huts behind them for native workers. Many native men walk or ride bicycles along the edge of the road, most in shirts, trousers, and hats but some with coloured blankets wrapped around them, and some with faces painted white.

As we drive, Jakes Ewer tells us something about himself and about Grahamstown. He and his wife, Griff, were born and educated in England. After war service and doctoral studies in entomology at Cambridge University, Jakes, along with Griff who is also a zoologist, came to Africa to teach at the University of Natal, and then to head the zoology department at Rhodes University, researching various aspects of insect physiology.

Grahamstown was founded as a British garrison in 1812, established to fend off African attacks against the Europeans, which continued until 1853; stone watchtowers still stand on hills surrounding the city as a memorial to its bloody past. The town's importance increased with the arrival of British settlers, but it still has only twenty-three thousand inhabitants, ten thousand of whom are white.

We reach Grahamstown in the early afternoon, a spic and span "English" town with twenty-four churches, a number of schools, and Rhodes University, with its seven-hundred preppy students all living in residence. The university, composed of many white buildings with red roofs built mostly after

World War II, is still so new that some of the residences are not yet in full use. Jakes drops off the education student and her suitcases at her dormitory, then stops at the Zoology Department to see what's up.

The department is spread over three small buildings. Jakes takes me to the Reprint Room in one of them where he has arranged a desk for my use. Around the desk are shelves stacked with reprints of articles written by faculty members that the secretary sends to anybody requesting them. In the corner is an electric heater because it's cold, both inside and out; I'm relieved to see it as I've only brought one sweater with me to Africa, assuming that I'd always be warm here. I'm to share my quarters with two polecats, which will have free run of the room. The polecats had been a nuisance in Anne Alexander's lab next door, their previous home, where they knocked over cages containing scorpions.

Anne is a doctoral student of about my own age, who will become my best friend at the university.

"How are you?" she says in a friendly voice, shaking my hand. "Did you have a good trip?"

"Wonderful, but I'm glad to be here."

"Let me introduce my roommates," she says, indicating several cages on tables beside her desk. "Here are my scorpions and here are the peripatus." She picks up one of the scorpions in her hand to show me.

"They don't usually sting," she says lightly as I draw back. "I've been stung a few times, but it isn't too bad. Right now I'm stimulating the end of a scorpion's leg with an electrical impulse to see which muscles twitch."

"Anne has just discovered how scorpions mate," Jakes says proudly. "Nobody figured this out before because they kept their specimens in cages without soil."

"Yes," Anne breaks in. "The male secretes a spermatophore which has to stand up in the earth. Then he grabs the female by the claws and dances her over the spermatophore which she sweeps up into her body."

"Isn't that romantic!" I say.

"Terribly," Anne agrees. "I've written a paper about it—something the world needs to know about." I look at the scorpion in her hand with new respect before she puts it back.

The peripatus, small, soft-bodied, pinkish invertebrates looking something like centipedes, thrill me too.

"I've never seen one before, but I wrote a paper on their gaits in fourth year," I say with excitement. "I got an A. Do you know about their different gaits?"

Zoology group at Rhodes University with, from the left, Willy, Anne Alexander, Jakes Ewer, and Griff Ewer (with pipe). Kit Cottrell is in the second row.

"Of course," Anne replies, although there must be few people in the world who do. "Peripatus aren't very common around here. These came from under a log up the hill."

We watch several moving slowly and, with gentle prodding by a pencil eraser, faster across the floor of their cage. (Little do I know that I'll eventually make gaits—those of the giraffe and other large mammals—the topic of my own doctoral research after I return to Canada. The study of gaits of animals may not have much practical use, but at least it doesn't hurt any of them.)

Outside Anne's room I catch a whiff of formaldehyde, reminiscent of any zoology department. So little is known scientifically about the animals of Africa that Griff, when out driving, stops to collect any roadkill she notices at the side of the road; even animal remains that are rotting and partly consumed by maggots are of interest because their skeletons can still be salvaged. To the dismay of her passengers, if she has any, she thrusts the carcasses into the boot and later transfers them to jars of preservative for future study.

The corridor which accommodates some of these jars is also lined with funnels, mice cages, skull displays, bookcases, empty boxes, and a bathtub holding several frogs waiting to be dissected. It also leads to a third room which belongs to two senior students, one working on pineapple soil and the other on ticks. Neither is there, but Anne, Jakes and I enter the lab anyway. Two rabbits and a white rat scurry away from the door. I crouch down to acknowledge the animals which, unused to so many sudden strangers, have retreated to the far corner.

"I used to have a pet rat called Ararat," I tell the rat in my rat-friendly voice, stooping down and holding out my hand toward her in a cordial gesture, but she isn't interested.

"We had a pet African porcupine as well," says Anne, "but he wandered away and was eaten by the cooks in the women's residence. We can't prove it, but there was a pile of quills beside a cooking pot and we never saw him again."

Jakes drives me to his house, where I'm given a small bedroom with four blankets on the bed to sleep in until I can find other accommodation. I meet Griff, who has short, black, bobbed hair and is pleasant and individualistic: unlike other women she wears slacks instead of skirts, smokes a pipe, sits on the curb to wait for a ride (which embarrasses her children), and swears when the occasion demands. She is devoted to drama, having just finished directing *A Midsummer Night's Dream,* and before that playing Agatha in T.S. Eliot's *The Family Reunion.* However, as a PhD in zoology, her first love is animals. As well as teaching, she is currently researching the paleontology of fossil pigs and the behaviour of her cats. She plans to watch which of the kittens in the female's next litter nurses on which teat, to see if they each have a favourite one. (She finds out that they do, each apparently locating its preferred teat by smell.) Later she will research and publish two important books, *Ethology of Mammals* in 1968, and *The Carnivores* in 1973.

Thinking about her pig research, Griff says,"Did you know that one article on pigs is written entirely in Afrikaans? Yet worldwide this means that virtually nobody else will be able to read it. I live in South Africa and even I can hardly understand it. Apparently it's easier to get grants if you publish in Afrikaans." She rolls her eyes, then changes the subject.

"Since you're going to be studying giraffe," she says, "I'll give you a letter of introduction to Dr. Leakey so you can look him up if you get to Nairobi. We've been on several field trips together. He's a wonderful person." Griff is obviously a woman of many parts.

Soon Anne arrives with Trevor, who teaches at Fort Hare, a nearby college established for Africans by Rhodes University, which Josiah Chinamano attended, as well as Nelson Mandela and Robert Mugabe. The five of us, later joined by the Ewer children—Bridget, fourteen, and Paddy, twelve—spend the afternoon chatting around the open-hearth fire. What is the chemical composition of the wing-cases of insects? Can an animal in a drought survive by drinking its own urine? One passionate discussion focuses on the good and bad points of the new film *Richard III,* during which we drink quantities of tea supplied by the Ewers's two "girls," Florence and Violet.

Because of the Ewers's liberal bias, these Xhosa women are treated with respect but, because of apartheid (pronounced *apart hate,* as if to reflect its reality), not like family. They live in rooms built onto the back of the house and do the Ewers's housework and all the cooking except for Saturday and Sunday dinners. There is little entertainment for them beyond their work because at 9:30 each evening, after a siren sounds, no natives are allowed on the street without a pass signed and dated by their employer stating where they are going, why, and for how long. Without such a pass, he or she goes to jail. According to apartheid law, black and white people can't sleep under the same roof at the same time, so Florence and Violet have to be locked out of the house after they've washed up the dinner dishes; Jakes or Griff has to get up at six in the morning to unlock the back door so they can come back in and make breakfast. If this is one of the best jobs for native women, as I've been told, what would the worst be like?

At the dinner table, where I'm seated at the foot, Violet serves stew, which is delicious, but because I'm nervous my knife slips when I try to cut a piece of meat and half the food dumps into my lap. It's hot and wet, but I'm wearing an Innis tartan skirt, which I imagine camouflages the spill. I'm too embarrassed to admit to this disaster. The others must surely notice the mishap but are too polite to draw attention to it.

As if nothing is wrong, we chat about giraffe and how I should go about researching them.

"Most of all I want to study their behaviour," I say, looking up earnestly while discreetly mopping up gravy from my skirt with my napkin, "but it would also be super to go to East Africa and see giraffe habitat there and a few more of the nine races. Each race has different kinds of spots, so it would be great to try and figure out why particular spots evolved in different areas."

"Have you written to Mr. Matthew, the farmer?" Jakes asks.

"Yes. Just before I left London I sent him a letter admitting that I was a girl, not a boy. I *had* to tell him sooner or later. He's supposed to write me here, care of you, to tell me what to do next." There's a pause while I surreptitiously spear a piece of carrot with my fork and slip it onto my plate.

"I hope it will be all right," Griff says, wrinkling her nose.

"His farm is in the eastern Transvaal near Kruger National Park, a thousand miles from here," Jakes says. "How will you get there?"

I hadn't yet thought of that. "By train?" I venture. Growing up with my family in Canada, trains had solved all our travel problems because we didn't have a car.

"No train goes there," Anne says.

"By bus?"

"There won't be a bus either."

"Oh." I hadn't imagined a country without lots of trains and buses.

"You'll have to buy a car," Griff announces. "That will be best. You'll need a car for watching giraffe anyway. They'll let a car come close to them, but not a person on foot."

"I'll be able to drive further north, too, to see giraffe in East Africa," I agree. I feel more comfortable now that the mess on my skirt has been largely returned to my plate.

"I don't know about that," Jakes demurs. "You have no idea how awful roads can be in Africa." We let the matter drop at that.

After dinner, when I've casually gone to my room and changed my skirt, we gather by the fire to drink coffee. Jakes reads aloud half of *The White Deer* by James Thurber, and we listen to Beethoven's *Fifth Symphony* on the record player, Anne or I jumping to our feet every few minutes to change the 78rpm records. I'm supremely content in this welcoming home.

On Sunday afternoon Jakes, Anne, and I drive out of town to collect scorpions for Anne's work. We stop opposite an invisible ostrich.

"I don't see him," I complain softly.

"By that big bush. About thirty yards back," Anne whispers.

"Big, fluffy plumage. Straight, pale neck," murmurs Jakes.

I peer without success at the scrubby trees and bushes where they're pointing. After a few minutes the large black and white bird turns and flounces away so I can finally admire him.

I'm equally inept at discovering other animals. Anne has only to turn over a few rocks to come up with a number of scorpions, some six inches long, which she puts into refrigerator containers for her research; these she places under the car in the shade until we're ready to leave. Jakes is pointing out dung beetles, tock beetles, saw bugs, firefly larvae, centipedes, millipedes, and termites. The others kindly pretend that I'm first to see several fat-tailed geckoes, which rush across the hot sand from one shading rock or bush to another, leaving behind a delicate track in the sand. Later, in a fenced paddock, we glimpse a small herd of springbok, the national animal, along with prickly pear cactus and century plants.

Before we return home we visit a *koppie* or steep rocky hill, where, at the summit, there's a colony of hyrax, called dassies or rock rabbits. Dassies are brown animals the size of a woodchuck and according to their anatomy, related to the rhinoceros and the elephant, unlikely as this seems. They have sweat glands on the soles of their feet, which enable them to clamber across perpendicular faces on the huge boulders among which they live. They peer down at us with interest until we begin to climb up; then a sentinel whistles an alarm, sending them into crevasses among the rocks. From the top of the *koppie,* among huge piles of dassie droppings, we see long, rolling, uninhabited hills stretching to the horizon in every direction.

"Dassies like to defecate only at certain places," Jakes says. "If you have a tame one, you can train it to perch on the toilet seat."

"You should study dassies," Anne suggests to me. "Hardly anything else is known about their behaviour. We could put up a blind where you could sit and watch them every day."

"I'd need a car to get here," I object.

"You'll be buying one soon."

"It would be good practice for watching giraffe," Jakes agrees. "To refine note-taking."

"I guess I could," I say doubtfully. I'm not keen, when dassies are so small and my true love giraffe so tall. Later I do spend several afternoons watching the creatures on this *koppie* from a deck chair, but I find the exercise futile. I can't tell one beast from another, often there are none at all visible above me, and when some are in view they spend most of their time either looking idly around or munching vegetation. Boring. Later I'm glad I didn't spend much time with them when I read that, in the Serengeti, a hyrax colony lives on the roof of the wildlife administration building at Seronera. There, if one could stomach the strong smell of hyrax urine, observing them would be a piece of cake. Really too easy, because if windows and doors were left open, dassies would sometimes take over the building, climbing onto beds and into cupboards, breaking into food stores, defecating, urinating, and even smashing mirrors by fighting with their own reflections.

The Monday after I arrive at Grahamstown, I'm thrilled, at last, to receive a pleasant letter from Ian but appalled to hear also from Mr. Matthew. To my horror, he writes that, since I'm not a man, he can't have me on his farm. There is no place for me to stay. Because his wife is in California, settling his daughter into a university there and visiting relatives, he lives alone; it would be improper for me to live in his house with him and the other white managers don't have room for me in their houses.

"What shall I do?" I wail at dinner.

"You'll just have to write him again and persuade him to take you," says Griff sensibly.

"He writes that he's fifty-seven years old. Surely there can't be anything improper when he's so old?"

"He's not *that* old," says Jakes, rather affronted. He and Griff exchange glances.

"Maybe you could camp out?" suggests Bridget.

"Or live in your car, when you get it?" says Paddy.

"There's probably a store nearby where you could buy food," Bridget argues.

"What about suggesting that you stay at a hospital? There must be one nearby," says Griff.

All these suggestions sound difficult. "Are there any wild giraffe near here?" I ask hopefully.

"No," says Jakes. "They were never this far south."

Desperate, I write immediately back to Mr. Matthew telling him how far I've come to study the giraffe (which he already knows), that he is my only hope (which seems to be the case), and that perhaps I can stay at a hospital if there is one nearby. I say that I will be happy to buy a tent and sleep in it, and work out of my car (which I don't say has yet to be bought).

Over the next days I hunt for the car which I'll need even if I can't go to Mr. Matthew's farm. With a car I'll at least be able to drive north and see giraffe in some of the game parks. After reading ads in the newspaper, I purchase from a garage a 1950 green Ford Prefect for £200, which is easy on oil and petrol, thank goodness, but passionate for water. The radiator needs to be refilled every twenty minutes when I'm driving. I call her Camelo, a shortened form of the specific name of the giraffe, *camelopardalis,* so named because it has spots like a leopard and is big like a camel. The Ewers and I celebrate my purchase before dinner with glasses of sherry.

To further celebrate the acquisition of Camelo, euphoric to have my own wheels for the first time in my life, I go for a drive on the "highway" to East London, concentrating to keep on the left rather than the right side of the road. As I struggle up a hill in second gear in the African location, the segregated area set aside for African workers, several men beside the road in torn shirts and pants make struggling motions with their arms and grin at me. I grin back, delighted at their attention. I pass a cart pulled by a span of six donkeys, and then a row of little girls carrying huge bundles of branches on their heads. When one waves shyly at me I immediately wave back, thrilled. Then I decide to wave to all the people I pass. When some wave back and smile, I'm overjoyed.

Opposite two small huts near the road I stop to take a picture. Immediately a man in brown shorts with a bead bracelet on each upper arm comes out to see what I'm doing. I ask if I may photograph his house, but he doesn't understand what I am saying. When I point to my camera, his face brightens and he says "Snaps." He calls to the rest of the family, his wife and small child and, apparently, his mother, who come out to join him. They all look pleased at my attention although also confused. What on earth am I up to? Why do I

An African family living along the East London to Grahamstown road.

want their photo? When I have finished I give the man sixpence, at which he nods his thanks. They all smile and wave as I drive away.

When I return along the road on my way back to Grahamstown, the family are outside and again wave happily to me. I wonder if they can survive on the rows of vegetables growing near the house. Or do the man or women work in Grahamstown? Native men can be labourers, messengers, garagemen, and policemen (who can arrest natives but not whites) in cities, but not store clerks or bank tellers.

That evening Jakes has bad news in a letter he has received from the Centre for Tsetse Research near Shinyanga in Tanganyika, to which he has written on my behalf. In his letter he had mentioned that I was female.

"It says they have no accommodation for guests," he states perusing the letter. "It says the nearest giraffe are twenty miles from the station over a very bad road and there aren't many of them. It says there are rhinos in the area and if one charges, there are only small thorn trees to climb for safety. And, of course, there are lots of tsetse flies."

"I gather they don't want me there," I comment drily.

"I think we can assume that. I'm sorry."

A week after my arrival, Anne and I move into the house of Dr. Pocock, a botany lecturer, to house-sit for her while she is away visiting her sister. For a month's stay we each pay £8 for our room, breakfast, dinner, and laundry; this will do wonders for my budget.

"Thank you so much for having me," I say to the Ewers on the morning I move out. "How can I ever thank you?"

"We'll think of something," laughs Jakes, a comment I'll soon regret.

Dr. Pocock leaves us in the care of Maggie, her Xhosa servant. Maggie is a widow who makes our breakfast and dinner every day, does all the dishes, cleans, and makes the beds. She isn't allowed to sit in any room except the kitchen (Anne doesn't see anything strange in this) so she passes her spare time behind the house, relaxing on a hard chair in the sun or in her tiny room, reading her Bible or the newspaper. Occasionally she has some friends to tea in the kitchen. She has separate rations of food (ample mealie meal or corn, and two pounds of butter, half a pound of margarine, and three pounds of meat weekly), because she isn't allowed to eat any of the wonderful meals she cooks for Anne and me. Why rations, I wonder? Is there a war on? She is supplied with clothing, paid £10 a month, and has two afternoons off a week when she visits her daughter in the native location outside Grahamstown, taking with her any of her own leftover food.

Sometimes I dry the dishes after Maggie has washed them, even though this makes her extremely nervous. She keeps muttering as she works, saying "I can do that," and "Don't you bother, now." I think of hierarchical arrangements. Her comments remind me of how upset women in Canada become if a male guest insists on clearing the table or helping with the washing up (although this almost never happens)—as if this is against the natural state of things. I want to show Maggie that she and I are equal but she knows we aren't, that unlike me she can lose her job at any time and be destitute. Much later I read a paper about baboons in which the author, Thelma Rowell, states that a hierarchy is set up among animals and maintained largely by the subordinate rather than the dominant individuals: "You cannot chase someone who doesn't flee," she writes. The learned insistence of inferiority, I think. I hope soon that women will *really* treat men as equals by insisting that they do their fair share of housework.

"Could we see your daughter sometime?" I ask Maggie when we are working together at the sink one evening.

"Yes, Miss Anne," she says. (She refers to us interchangeably as Miss Anne and the Other Miss Anne.) "Margaret lives with my brother and his wife and daughter. She's five."

"Could we drive out some afternoon and see her?"

"Yes. I'd like that," she says with a big smile.

Anne and I soon arrange this, driving out in Camelo on a Friday for a surprise extra visit. Maggie points out where her friends live as we cruise through the location, past row after row of similar huts spaced closely together. We pass women in faded frocks walking along with handbags, shopping bags, and in one instance a suitcase, balanced on nests of cloths on their heads. Some have a wide cloth wrapped tightly around their middle in which a baby rides on their back.

"There's a man with his face painted orange! I wonder why!" I exclaim.

Anne, startled, looks to the left where I am pointing. "That's not a man, that's a boy," she says, shrugging her shoulders. "Some ritual, I suppose."

Maggie doesn't say anything, although she must know why.

When we reach Maggie's street, C Street, about twenty small children rush to greet her, all shouting excitedly; some are dressed in tatters while others have no clothes on at all. Margaret, her head a mass of tight curls, throws herself on her mother and hugs her so hard that Maggie can't walk.

When Maggie finally breaks loose from her daughter she proudly shows us her brother's house, which he rents from the government. It's about ten by twenty feet, with a curtain strung across the middle.

"This side is for sleeping and the other for living in during the daytime," she explains. "There isn't electricity, but we have this oil lamp, which works all right. We get water from a tap down the street."

"It's quite close," says Anne, peering at the tap through the door.

"Sometimes, in a drought, there's no water at all, but then government trucks bring water in."

"Good," says Anne.

"The walls and floor are solid—made of cow dung. My brother has whitewashed the outside walls and covered the roof with corrugated iron to keep out the rain. Here, at the back of the house, is the garden where we grow mealies."

I'm aghast at the primitive conditions, but try desperately to think of something positive to say because Maggie is so proud of the house.

Not Anne. "This is wonderful! What a lovely home, and what lovely pictures on the wall," she tells Maggie enthusiastically, referring to several religious scenes stuck up with tape.

"Thank you," says Maggie. "I'm glad Margaret can live so close to me."

"You're really lucky. The natives that live in Natal where I come from are much worse off." I can hardly believe my ears, comparing in my mind this tiny family house with Dr. Pocock's splendid, eight-room mansion. Anne is liberal, but because she was born in South Africa she is conditioned to think of European and native standards as completely different. Would I think like her if I were also born in South Africa? Almost certainly. It's an awful thought. Even worse, Maggie is pleased at this compliment.

"Yes, Miss Anne," she says. I long for a look of irony on her face, but see none.

Although Maggie can read and so is more educated than most Africans, she is still superstitious. Later Anne tells me that when one of her relatives was sick, she asked Anne for money to buy a goat to be slaughtered so the woman would recover. We look at each other in amusement, rolling our eyes. Thinking back at our reaction decades later, I'm embarrassed to recall our response; I remember that four years before I even met Maggie, when my father lay dying of cancer in Toronto, on every rainy day on the way to or from the university I searched out rainbow circles on the road from oil spills, stepping on each so that I could make a fervent wish to God: "Please, please dear God, don't let him die!" Maggie was probably equally desperate, but I hope her wish facilitated by the poor goat was better rewarded than mine.

# 3

# Rhodes University

My month at Rhodes University spent waiting for a positive response from Mr. Matthew gives me a chance to compare it with the University of Toronto, a much larger institution that I know well because it employed my mother and father, and educated my brothers and sister and me. The major contrast between the two is the open racism practised at Rhodes, which benefits all of the students and faculty (all white, of course) at the same time as it desensitizes them to human justice. Yet, despite this, so much is the same at the two universities—entertainment, rivalries, small talk.

I spend most days in the small zoology library searching for information about giraffe. One reference discusses the first individual to reach Paris, in 1827, after walking all the way from Marseilles, where the mayor had led the animal on a public stroll each day to train for this march. The animal caused a sensation, needless to say, in every town through which it passed. Other scientific journals contain anatomical notes on giraffe, mostly of individuals who died in zoos. There are lengthy discussions on the presence or absence of a gall bladder, on the exact arrangement of blood vessels connected with the aortic arch, and on the unusually wide structure of the lower canine tooth, which enables a paleontologist to recognize a member of the giraffe family from that tooth alone. Odd, I think. When I visualize giraffe, I think of them as living animals wandering freely on African savannahs, not as soulless assortments of unusual organs, blood vessels, and teeth.

When I've run out of giraffe material I start on books and articles dealing with Africa and Africans, some of which touch on giraffe behaviour in passing; these were written by "sportsmen," missionaries, or explorers. The hunters describe details of how Arabs kill a giraffe by severing the hamstring muscle of a hind leg so it can't run, while the others mention where giraffe live and how tall/fast/timid/voiceless they are. The material makes me keener than ever to venture into the wild to observe giraffe and their habitat for myself.

Of course, the giraffe was not the only common animal whose life history was ignored up to the 1950s. The life history of virtually all animals was unknown. Zoologists were busy in the first half of the past century naming and cataloguing species, and studying animal physiology and anatomy in the laboratory. It was easier to measure the leg of a dead giraffe than to trail a live animal through a thorn thicket to watch the way the leg moved. It was easier to sit and measure a giraffe's canine tooth than to tramp over hot fields documenting the diet that affected the tooth's evolution. The reason why we know little about giraffe is clear, and just as obvious, I think, is the need for me to rectify this gap in our knowledge.

The zoology department takes a break each morning at 10:30 and each afternoon at 3:30 when Jakes, Griff, the three lecturers, five senior students, and I congregate for tea. A tray complete with steaming tea, milk, sugar, cookies, spoons, cups, and saucers appears as if by magic on the verandah when the day is sunny and warm, or in one of the labs when it's cold. When we finish our break, the dirty dishes disappear again. It is several weeks before I see Willy, the middle-aged, portly zoology department's "boy," who makes this splendid ritual happen.

The conversation at tea is general, focusing one day on whether Christianity applies to Africans.

"Of course it does," says one graduate student. "The Bible specifically states that natives are chosen by God to be hewers of wood and drawers of water."

"I think religion is good for natives because it helps them put up with their dismal lives," says another.

"I think religion is above and beyond Africans, because they have no real human potential the way whites do," adds a third.

I sit in stunned silence at these comments.

Our daily lunches at the women's refectory make me realize how easy it is to be sucked into an apartheid-like system of hierarchy. Anne insists we eat at the head table, which stretches the width of the large dining room at one end, and is set on a dias several feet higher than the rest of the tables and chairs allotted to the hundreds of women undergraduates, all white. Anne, along with her guests, is eligible for this honour because she will be a don next year.

"The head table is set up *for* faculty," she explains. "The students go through the cafeteria line, but we're served by waiters."

"It's so inegalitarian," I object. "Don't you feel like the French queens long ago, who indulged the populace by allowing them to shuffle by and admire them as they scoffed down their meals?"

"It works out all right. You'll get used to it," Anne insists, a little testily I think. The waiter brings bread, butter, steak, corn fritters, beets, cabbage, and tomato salad. When I take an orange for dessert, he rushes over with a finger bowl so I can wash my sticky hands after dissecting the fruit.

"The waiters aren't allowed to enrol here," I comment. "All the students are white."

"Africans go to Fort Hare College, which is set up just for them. What's wrong with that?" She thought for a moment. "Actually, the government is shutting down parts of Fort Hare so I guess that doesn't necessarily work after all."

I often sit beside the Dean of Women, a matronly person of a certain age who is interested in her cat, not Fort Hare. "She'll be having kittens any day now. Isn't it exciting? It will be wonderful to have young things around my place. I'm not married, so the kittens will help keep me young." She looks at me questioningly. Am I to say nonsense, she's youth itself, or that the kittens are certainly needed to keep her young?

The student body pays no attention to us superior beings at the head table—the girls sit in groups chatting and munching down the food they carry on trays to a table. Nor do the waiters seem to think that anything is wrong.

Gradually, after a few weeks, I no longer notice the discrepancy in treatment afforded blacks and whites, and to undergraduates and "faculty" such as ourselves. The difference disappears for me as it had far earlier for Anne and everyone else born in South Africa. It seems normal that proven scholars should be served at the head table by a black waiter without schooling, and that we, with our dainty finger bowls, should look down upon the less-educated girls with their thick, china dishes. At least the waiter has a job, and the undergraduates chattering away below us seem happy enough.

One day I meet Jakes in the hall. "I've thought of a way," he says, startling me.

"A way?" I ask, wondering what he's talking about.

"A way for you to thank us, as you said you wanted to do. Why don't you give a talk to my second year zoology class on the genetics of mice—my master's research topic—giraffe, Canada, or any subject you like? They're amazingly provincial so I like to expose them to all sorts of ideas and people."

"They'll love your Canadian accent," says Anne, who is just coming out of her lab.

"I suppose so," I say doubtfully. At my high school I had gloried in being head of one of the sports groups, but during my university years in the early 1950s my self-confidence had been eroded by sexism. Boys but not girls were hired for well-paying field work. Girls paid the same tuition as boys, but were relegated to inferior sports facilities. One male graduate student insisted that my body build was ideal for having babies rather than for being a zoologist and messing about with animals. However, I can't possibly say "no" to Jakes, who has done so much for me; anyway, this will be a good chance to overcome my fear.

"How many students are in the class?" I ask.

"About twelve. Nice kids. You'll like them." I feel my stomach clench in nervousness.

I find it amazing that I've come half way around the world, yet so many social events at Rhodes University are the same as they are at the University of Toronto. In both places movies play a major role in social life in this pre-television era. The difference is that at Rhodes, movies (or "bioscopes" as they're called) are more formal and, of course, segregated by race. The seats are all reserved, with ushers to guide you to your row and patrons well dressed in blazers or suits, or full skirts and blouses—the university is not called a "marriage market" for nothing. Non-whites are only allowed in one of the houses, and only in the last two rows in the gallery; Indians settle there, but not natives because of the evening curfew. After the first twenty minutes of short films, there is an intermission before the main feature begins, during which the audience, mostly couples, rushes out to the nearest restaurant for a soft drink.

There are also lots of dances, which is how I get to know Neville Meyers; he is doing graduate work on ants, studying why one of their major food sources, the "honey dew" excreted by aphids, is more abundant at some hours of the day than at others. When he asks me to his residence ball, it doesn't occur to me to refuse because I have an "understanding" of sorts with Ian. Indeed all the time I am in Africa I never mention Ian to anyone. My secret life? He hardly seems real in this exciting new world. I tell Neville that I would love to go.

The ball is a glorious dance with fast music and many waltzes where the men wear tuxedos. I wear my only dress (yellow) and Anne's velvet cloak. Between sets, couples relax on the lawn. Neville points out the Men's Warden, an ardent Nationalist who spent much of the last war in a concentration camp because of his Nazi sympathies; indeed, during the war the vote to support the Allied side was only narrowly won by the government. To my horror, the Warden asks me to dance, and I accept, not knowing how to refuse.

The servant Maggie and Anne Alexander outside Maggie's house.

What can I say to him? I desperately dream up light conversation. How strange that I have never been physically close to any of my girlfriends, yet here I am pressed against the chest of a man I've never met before, and in this case a Nazi!

On another evening, Neville and I go to a dance at Oriel Hall, where each weekday Anne and I eat lunch. I again don my yellow dress while Anne repossesses her velvet cloak. I waltz with Jakes, who asks me not to mention his drunken parties when I write my book about Africa. During a Paul Jones set I win the female prize for dancing with the "mystery man," so designated by the disc jockey, whose name I never learn. It's a small, china Bambi statue, which I give to Bridget the next day.

At both the University of Toronto and Rhodes, the students, virtually all middle-class whites, enjoy themselves in groups as well as in pairs. One day Kit Cottrell, Neville, Anne, and I drive to the ocean to spend the day. It's so cold and windy that we have the beach to ourselves, a huge expanse of sandy coves and rocky promontories. The men swim in the icy water while Anne and I check out the ocean fauna on the tidal plain. For lunch we build a fire, over which we cook sausages and steak to eat with bread; we've scarcely finished preparing the meal when an ocean wave sweeps in, drowning the flames. After lunch we play leapfrog, tag, and bowls using oranges. Within thirty years, bodies of African men assassinated by the police will be discarded along this beach east of Port Elizabeth.

On another evening the four of us drive to the top of Mountain Drive overlooking the city, where we have another sausage roast, this time with

Maggie's daughter, Margaret.

sherry. I don't know or understand the Latin student hymn "Gaudeamus Igatur," nor the songs they sing and jokes they make in Afrikaans; they speak far better in this guttural language, that of the Nationalist Party in power, than Anglo-Canadian students speak French. I do know all the English songs—"Roll Out the Barrel," "It's a Long Way to Tipperary," "On Ilkla Moor Baht 'at," "Good-Night Irene." What a small world it is!

After the sherry has taken full effect, the others discuss whether it's possible to tell from their behaviour if Jakes and Griff have had sex the night before. Opinions are divided. I pass on this question. We end the merry gathering by rolling down a nearby hill. Although Neville and I land in a close tangle of legs and arms, I refer to our friendship as platonic—maybe I think that because I haven't yet kissed him, there is no disloyalty to Ian? Neville writes me later to say that platonic doesn't describe the situation—that plutonic, as in plutonium, is more apt. We arrange to meet when I return to Rhodes, but by that time he is no longer there. I never hear from him again. Ships that pass in the night.

The world seems even smaller when Professor Edgar McInnis, a historian from the University of Toronto, comes to Rhodes University to give a talk about Canada. When he sees me in the audience wearing a University of Toronto blazer, he comes up to greet me.

"You knew my late father," I say.

He looks at me quizzically. "He was...?"

"Harold Innis, the economic historian."

"Of course!" Immediately he and I are old friends. He introduces me to the Vice-Chancellor, who performs the role of president at Rhodes University. Dr. Alty taught physics at the University of Saskatchewan for ten years, so he is interested in Canada too.

"I met your father, and mother, and sister Mary in England," he says, equally amazed at this coincidence. His wife invites Anne and me to tea the next week at their beautiful house on campus. We go in our best finery (my yellow dress again), wearing white gloves; the four of us discuss education, South Africa, and Canada while a native man dressed in white passes around tea, dainty sandwiches, and cookies.

That night the Ewers entertain Anne and me at dinner, where we talk about this social occasion and about university life in general, a scattered type of conversation undoubtedly similar to that taking place in Toronto, albeit on different topics. While we chat, Bridget and Paddy play Monopoly on the floor at our feet and read bits of *Dr. Doolittle* when they get bored—exactly how my siblings and I amused ourselves while we were growing up. They often badger their parents to buy a radio so they can listen to programs their friends discuss, just as in Canada children are beginning to beg for television sets.

One topic which seems almost mandatory because it is so common is that of books banned by the government.

"I've heard that twenty-three hundred have been banned!" says Anne.

"Apparently the government was embarrassed when people complained about *Black Beauty* being unobtainable on racial grounds, so they took it off the banned list," says Griff as we all laugh.

The government rescinded the decision to ban *Before We Go to Bed* as well," says Jakes. "It's not about pornography but a book of children's prayers."

"At the University of Toronto we used to laugh about books banned by the Catholic Church. The list of forbidden titles in Quebec was much longer than in Ontario, so some Quebec Catholics used to hole up in Toronto hotels to read books *verboten* in their province," I say. The others nod and laugh.

"I've just read *Old Four Legs,*" I say, changing the subject. This is a recent book about the coelacanth, a huge deep-sea fish until recently believed extinct, which Professor Smith has had transported to the campus and which I have seen reposing in a six-foot-long tin coffin awash in dirty formalin.

"I saw Smith on campus today. I expected his head to be as big as a balloon with the hype about his book," say Jakes.

"I really doubt that when he first saw the coelacanth he shouted, 'A coelacanth, supposedly extinct for millions of years!'" adds Griff rather snidely.

"I heard that nobody knew *what* it was when the first one was caught," says Anne. "It was Miss Latimer at the East London Museum who figured it out. She gets little credit, of course."

I'm reminded of the many heated discussions I've heard at the University of Toronto about Frederick Banting and Charles Best discovering insulin for the treatment of diabetes: was it fair that the Nobel Prize went to Banting and another man, but not Best? Certainly Banting and Best didn't think so. University passions over personal achievements must be the same the world over.

The day after this diversion Jakes finds me in the library, nursing a swollen finger bitten by one of the loose zoology polecats. I was reading a book on my desk at the time with my hand hanging down, so perhaps it was the animal's astute way of attracting my attention.

"The speaker for the zoology class today is sick," he says. "Can you give your talk this afternoon? You don't have to say much—just answer their questions about Canada if you like."

I gulp, my stomach giving a lurch. "I guess so," I say.

"The class meets at two o'clock in the small lecture room. I'll introduce you, then leave you to it."

I've prepared notes for a talk on what is then called "racialism." It seems to me almost criminal that the students at Rhodes are apathetic about politics (although in Canada I was too), and to the plight of the natives whose barren lives are being increasingly restricted by apartheid.

The students must know about native problems: every few weeks there are pictures in the papers of Black Sash women—women standing silently wearing black sashes outside government meetings to indicate that the government is flouting the South African constitution in its persecution of natives. Huge native protest rallies and "actions" organized by the African National Congress and the Pan-Africanist Congress will soon have these groups banned and Nelson Mandela imprisoned for twenty-seven years. There is already so much political unrest that, in a few months, one hundred and fifty-six people of all races will be arrested to face trial for treason, punishable by death. University is a time for questioning, for discussing anything at all, I idealistically believe. It's a time for planning and preparing to change the world for the better. I imagine myself lighting a fire under these students which will rouse them to take up the cause of the downtrodden Africans!

The focus of my talk will be a comparison of race problems in South Africa with those in the southern United States. In both countries there is segregation of blacks and whites, killings, and attacks on black schoolchildren. As a Canadian, I feel (undoubtedly unjustified) morally superior to both countries. First, I'll tell them how badly blacks were treated as slaves in the United States, pointing out that even after slavery ended blacks still faced segregation, which led to inferior housing, poor education, and dismal jobs. I'll talk about lynching and the Ku Klux Klan. Then I'll show them that things are slowly getting better by talking about Paul Robeson, Louis Armstrong, Marian Anderson, and the recent confrontations for human rights.

In contrast to this hopeful situation, I'll emphasize how discrimination against native South Africans is *increasing* as apartheid is implemented, leaving the natives in despair and Europeans with an exaggerated sense of their own importance. Like a zealot, I'll force them to think about racism and the harm it causes everyone. I have a list of books for them to read that I'll write on the blackboard: *Black Boy* by Richard Wright, *The Dark Child* by Camara Laye, *The Colour Problem* by Anthony Richmond, and the newly published *Naught for Your Comfort* by Father Trevor Huddleston.

Armed with moral indignation, I'm only slightly nervous facing the class after Jakes has withdrawn. The students, sprawled in a relaxed manner on chairs set up in rows, regard me with interest.

"I thought we should look at the treatment of natives in South Africa," I begin. "I'm from Canada and so see the situation from a fresh perspective." I look up brightly.

A boy at the back raises his hand. "We're not allowed to discuss politics at Rhodes. It's against the law," he says matter-of-factly.

"Not allowed?" I respond, astonished.

"No, it's not allowed," agrees the girl next to him. "That sort of talk leads to trouble. The Dean doesn't allow it."

I'm too stunned to come up with my philosophy of an ideal university, where *everything* should be discussable. There is a long silence. I look at my notes. What on earth else can I talk about?

The same girl finally breaks the silence. "Is it true everyone carries guns in America?" she asks.

Thanks to their questions we talk about crime; how Canada is not the same as the United States; snow, which they have never seen; French versus English tensions in Canada; and why the reserve system for Canadian Red Indians (as they call them) is not the same as apartheid in South Africa (although it is probably more similar than I choose to concede). At the end of the period I walk out of the room, completely deflated.

"How did it go?" asks Jakes, who is waiting outside the room.

"It went," is all I can come up with. I hope this is the worst "lecture" I will ever give.

A few days later Griff phones me at Dr. Pocock's house at dinnertime to say that a telegram has arrived from Mr. Matthew. Anne and I walk to the Ewers's house in trepidation. What could a telegram mean? Better or worse news than a letter?

"Here it is," says Bridget, handing me the telegram. "Cross your fingers."

I take the envelope gingerly. "Please say yes," I tell it. "Please say yes."

When I tear it open, this is what the telegram says.

"I can go!" I whoop. "He says to come! I can't live at the hospital, but I may be able to stay at a neighbours, or even at the farm itself. At Fleur de Lys Farm! Hurrah!"

The others are delighted. "Well done!" says Jakes.

"I'm so glad!" Anne and Griff exclaim.

Even Florence and Violet look pleased when they realize what is happening. Mr. Matthew tells me later that he decided that anyone who had come to South Africa from as far away as Canada should at least be given a chance to carry out their dream.

On my last night in Grahamstown, the Ewers give one of their unique parties that has been planned for weeks. I bring a large box of chocolates and a copy of the new book *You Are Wrong, Father Huddleston* as a present for the Ewers to thank them for all their kindness. Griff, who left the room as Jakes was opening the gifts, returns and picks up the book.

"What a waste of money," she says, putting it down again with distaste. She finds its thesis, that apartheid is a commendable form of government, disgusting.

"Anne brought it to us as a present," Jakes says drily.

"Oh," says Griff.

When everyone has arrived, we have a play-reading of Shakespeare's *Henry the Fourth, Part One,* with Jakes taking the role of Falstaff and Griff of Glendower. As a newcomer, I read the parts of an assortment of messengers, servants, and attendants, warned by a jab in my back from Jakes, at whose feet I'm sitting, when one of my lines is coming up. Then Jakes sings a hearty rendition, in the spirit of Shakespearean times, of "With Her Head Tucked underneath Her Arm," after which we feast on wine, beer, and buns with cheese.

Griff concludes the evening by reading a News Flash she has written (although hardly recent) entitled "Scientist Savaged by Polecat."

We deeply regret to have to report that during the course of her studies the scientist was the victim of an unprovoked act of aggression on the part of a specimen of *Poecilogale albinucha* (Gray). No official statement has, as yet, been issued, but it is believed that the aforesaid [animal] had, during his sojourn in the Zoology Department, become infected with the prevailing zeal for experiment, and, never having tasted Canadian before, decided that he would "try anything once." We understand that no second attack has been made.

We all laugh at this, but later I learn that a mongoose Griff found and brought into the lab had rabies, which meant she had to undergo an extensive series of preventative and painful shots. Thereafter, all the research students at Rhodes were ordered to be immunized against rabies every six months.

Jakes puts on a record of Beethoven, and as the music swells, I reflect on university life. I went to a large university with high standards, but no professors treated students as well as Jakes and Griff have treated me and theirs. They have a human touch. The faculty and libraries are small at Rhodes, but there are more important things in a university. The Ewers are devoted not only to their own research and teaching but to making sure their students receive the best education possible. Unfortunately for the university, the government policy of racism is so abhorrent to them that, a few years later, they leave Grahamstown to move to the University of Ghana. Jakes will later write an unpublished memoir about life and education at both universities.

By 2005, of course, Rhodes University has been transformed; it is captivating on the Internet to see black and white students studying and socializing side by side. Jakes and Griff would be thrilled at this.

As we rouse ourselves to go home after the music dies away, I know I shall miss the kindness of all my new friends at Grahamstown. But I've been at Rhodes for a month, it is now the end of August 1956, and I'm ready to focus on live giraffe in the wild.

# 4

## Driving to Giraffeland

I set my alarm for five in the morning, prepared to sneak off in Camelo without waking anyone. However both Anne and Dr. Pocock, who had returned to Grahamstown and her house the day before, insist on getting up, too, to see me off. Dr. Pocock presents me with a large packed lunch to eat on the road. By 5:45, when it's still dark, I'm ready to leave.

"Don't forget to write," Anne says. "And give my love to the giraffe."

"Stay with me when you come back to Grahamstown, if that fits in with your plans," Dr. Pocock kindly offers as I climb into Camelo.

"Thanks again for everything," I call from the car.

The Ewers had traced the best route for me to take on my new Automobile Association road map of South Africa. I should drive north to Johannesburg, east to Nelspruit, then north to Fleur de Lys and the giraffe. It sounded simple, but seemed anything but when I stared at the thousands of names and the roads criss-crossing the large map.

"It looks a long way," I gulped, gazing at the huge distance and the scores of towns that I'll have to pass through.

"A thousand miles," Griff said.

"The roads aren't bad though," said Jakes. "Until you get to Nelspruit, anyway."

"Are they all paved?" I asked.

"Of course not. This is Africa."

"Klaserie, the town nearest Fleur de Lys, isn't even on the map," Griff pointed out helpfully.

"Oh," I said.

"It'll take you four days in Camelo," said Griff, "but there are hotels and inns to stay in along the way. You should be fine."

Forty-five years later, when I read Jakes's memoir, I realize they didn't mention their other concern—that since Mr. Matthew's wife was out of the country for the year, my arrival and stay at Fleur de Lys meant that "the Veld might be scandalized!"

Griff had said that it would take me four days to reach Fleur de Lys, but she underestimated my anxiety to see the giraffe. I decide to drive each day from dawn to dusk, sustained in part by Dr. Pocock's sandwiches, and try to get there sooner. In theory, driving steadily without stopping is a poor way to sightsee, but as Camelo balks at speeds above fifty miles an hour even on paved roads and needs water refills every twenty minutes or so, I have ample opportunity to look about.

Travelling northeast, I drive first through an African reserve, the pasture-land dotted by neat round huts with thatched roofs where families still sleep. A single Xhosa man clad in a blanket stands watch by cattle fenced in their *kraal* by prickly sisal plants and rough planks. The occasional barking of a dog breaks the stillness of the fresh cool air. As I approach Kingwilliamstown, the Xhosa dwellings are no longer *rondavels* spread across rolling green hills, but deformed tin shacks, crowded together on rough land covered with thorn scrub and aloes; it's as if the town has cast a blight over everything around it. In this derelict place, only twenty years in the future, Steve Biko, as leader of the anti-government Black Consciousness Movement, will be forced to spend his last years of life under house arrest, although still able to organize the movement and receive visitors, one at a time, from all over Africa and the world. In 1977, while in custody, Biko will be brutally tortured and murdered by South African police.

Beyond Kingwilliamstown I drive along a dirt road through the Transkei Native Reserve; again there is pastureland for cattle, but erosion is so heavy that the bare earth colours the land red instead of green, and dozens of gullies crisscross the surface, funnelling top soil away to the sea. There are few people about in the Transkei, mostly men and a few ragged children who stare at Camelo, but for the next hundred miles beyond there is no one at all on foot. The road winds up and around and down eroded, round-topped hills.

Camelo needs constant attention. After chugging up steep hills, I stop beside rocks or an occasional fir plantation to allow her engine time to cool as I refill her leaky radiator. More rarely I pause to buy petrol at a tiny petrol station; in this deserted region there's a petrol station every thirty miles or so, which seems excessive for the small amount of traffic.

Camelo on the way to Fleur de Lys, with cooling water bag on her radiator.

I stop briefly to eat a sandwich at Aliwal North, parking under a tree shading the sleepy main street. Before me is the Orange River, marking the boundary of Cape Province—which I'm leaving—and Orange Free State (the "Free" representing the historic Boers escaping from the British), which I am about to enter. Early explorers encountered giraffe north of this river, but never south. When I studied geography at school, I had imagined the Orange River to be a mighty boundary between English and Afrikaner colonies, not the sluggish, muddy stream I see before me. It doesn't empty into the nearby Indian Ocean as one might expect, but winds its way across the huge continent to the Atlantic Ocean.

I drive north with the mountains of Basutoland to my right. Toward Bloemfontein, the capital of the province, the countryside becomes more lush. First only sheep wander the hills; then the sheep give way to cows; finally, the cattle largely are replaced by expansive stud farms for hundreds of thoroughbred horses grazing on thick, green lawns of grass. The tails of horses swishing back and forth to chase off flies do so more gracefully than those of cows.

Bloemfontein proclaims itself to be a centre for robots, with signs everywhere: "Watch out for robots." I'm willing to watch out for them, but what are they? Finally I decide that a robot must be a stop light.

I had promised myself, for safety reasons, to stop driving before dark, but don't do so until I reach the dusty town of Winburg, where there is a small hotel looking rather like an overgrown bungalow. Only a dark-skinned waiter notices my arrival about nine o'clock. He is very friendly as I sit down to dinner.

"You from Grahamstown?" he asks eagerly in broken English. He has seen Camelo's licence plate through the hotel window, which indicates where I'm from.

"Yes," I answer. "Do you know Grahamstown?"

"Yes," he answers happily. "I been there."

"Did you like it?"

"Yes. Very much."

We smile at each other often after this, united by a tenuous link to a city hundreds of miles away where we would never have spoken to one another.

Next morning, after breakfast at six o'clock, Camelo and I continue our progress steadily northward, now between large cultivated farms. I enter the province of Transvaal shortly before Johannesburg, where Camelo and I struggle through the centre of town at noon. The traffic is impressive. It's been months since I've been in a double line of cars waiting for a stop light to turn green. I idle behind an African cyclist on a bicycle labelled "Jet Laundry." Thirty miles on I bypass Pretoria, drink a Coca Cola for lunch to save time, and turn east.

The paved highway to Nelspruit, three hundred miles away, is superb. Often it unfolds for ten miles at a stretch without a single bend. Sheep and cattle dot long, rolling hills; fields of swaying wheat and oats blanket others. The Africans in this region are Ndebeles, living in square houses surrounded by rectangular walls painted in brightly coloured geometric designs. The men and women both wear several coloured bead necklaces, each at least three inches in diameter, equally sturdy anklets, and little else.

Camelo responds eagerly to the wonderful road, rushing down hills at sixty miles an hour. On one such spree, on the bridge at the bottom of the hill, a truck going the other way crosses over the centre line and passes so close to Camelo that the tarpaulin strapped over its load bangs against her side. I narrowly avoid crashing into the bridge rail to escape a collision. If the driver was trying to scare me, he certainly succeeded. My heart beats so loudly and my foot shakes so badly that I draw over to the shoulder of the road and stop to recuperate.

Other vehicles are courteous. As the road leaves the pastureland of the highveld near Belfast to wind down the Drakensburg Range toward the subtropical lowveld of the eastern Transvaal, I follow an open truck loaded with African labourers. They're in high spirits, laughing and pointing at Camelo and poking each other. They help me pass their cumbersome vehicle by peering ahead and making wild motions to keep me back when they feel it's unsafe to pass, either because of the traffic or a bend in the road. They finally urge me on, with big grins on their faces, when it is clear ahead.

Human contact, I think happily as I speed around their truck. (Although Camelo and I are well-suited, I am beginning to feel that I've been driving in her continuously for a week instead of two days.) I enjoy the truck interlude so much that, ten miles later, I let the truck pass me. The men look startled to see me behind them again so soon, but they gamely scout ahead for me as before. Disappointingly, the truck and its occupants turn off on a side road before I can overtake them a second time. The Africans and I wave each other goodbye.

At Nelspruit, the citrus centre of the lowveld, I pause for dinner—two dishes of ice cream and a glass of milk. It will soon be dark and I know I should stop for the night, but I'm too wound up to do this. Only eighty more miles to the giraffe, I say to myself. And if I don't carry on, the telegram I sent Mr. Matthew saying I was on my way will worry him if he expects me tonight.

I start out on a road heading north, but stop at the first petrol station beyond Nelspruit to fill up and enquire about the route.

"Is this the road to Klaserie?" I ask.

"Never heard of it," says the man at the pump.

"Klaserie's near Fleur de Lys Farm," I clarify.

The man shrugs. "Don't know. Is it north of here?"

"Yes. Near Acornhoek."

The man rolls his eyes. "There's only one road north," he says, "so I suppose you'll hit it somewhere."

The road is paved for the first few miles, but after that, as the sun sets, it becomes a wash-boarded, gravel nightmare. Camelo can never exceed twenty-five miles an hour if we're not to be completely shaken up. Even at lower speeds I clench my jaw and tense myself against the car's shudders. There is no moon, so my headlights beam eerily into the jungle of darkness before me as though I'm in a tunnel. The only sound is the steady throb of Camelo's engine. By now I've been driving steadily for fourteen hours, so the noise of the car putting along in the night, shuddering and lurching as it hits yet another set of washboards, is drilled into my brain.

My mind is so numbed with fatigue that I'm startled at intervals to encounter a donkey cart or a native walking, surprised that other people inhabit this desolate spot. Twice my headlights pick out bloated puff adders lying at the side of the road, lured there earlier by the sun's heat. Once the hills around me are lit up by grass fires creeping slowly across the veld. I meet another car which whirls around a bend like a lighted phantom, passes me, and is immediately gone again.

An hour after I pass this car, and most of that since I've seen a human being, I figure that I must be nearly at Fleur de Lys, judging from my rough

The road to Fleur de Lys with the farm buildings visible
below the Drakensburg Mountain range.

map. I've been driving for three hours since Nelspruit. I stop at one of the few lighted houses on the road to ask for directions, a building that turns out to be part of a mission at Arthur's Seat, also not on the map. The white man who answers my knock gazes at me in amazement; I'm sure few strange women call there alone after dark. I stare bleakly at him, hoping improbably that I am already at Klaserie. I'm not.

"Follow the road for ten miles, then turn left. Fleur de Lys is the second road left after you cross the railway track. Drive along that road for half a mile to reach the farm buildings."

"Thank you." At least he's heard of Fleur de Lys.

The man closes the door softly as I turn away, still puzzled by my appearance. In a country where Europeans are warned never to drive alone, he obviously found the sudden apparition of a white woman in the darkness difficult to credit.

From now on the washboard surface of the road becomes even worse. I slow to fifteen miles an hour, then to ten to try to prevent Camelo's convulsive shuddering. Even so, she must feel the strain because, when I'm about five miles from Fleur de Lys, her engine chokes, dies, and refuses to start again. The only sight caught in the car headlights is the tunnel ahead—the brown dirt road and the dark bushes lining either side.

Damn. I try to start the motor. The engine growls but doesn't catch. I try again but the noise is fainter. Then weaker yet. No hope. I switch off the car lights. The silence is devastating. I'm stranded on a moonless night in the middle of nowhere. My mind begins to work overtime. I think of the newspaper report of the leopard that attacked a man near here two weeks before. I remember the poisonous puff adders that lay on the road. And what about pythons? Mambas? Cobras? And lions, of course. I recall the drunken natives who play a part in some books on Africa. I'm terrified. At that moment I would gladly have returned to Canada to start my adventure all over again.

I force myself to calm down. I can do either of two things. I can stay where I am for the night, or walk in the dark to the farm. The first alternative is the more sensible. With the car doors locked nothing can harm me, unless a lion or an elephant or people make a sustained assault on Camelo. This is unlikely. Why would they? I settle down to wait out the night, wrapped in stillness.

Minutes pass. I put my legs up on the seat beside me and take a deep breath. But my mind won't rest. I imagine a native man coming up to Camelo. Peering in. Tapping on the window. Probably gesticulating. Me looking back in the dark, shrugging my shoulders, trying to smile although he won't be able to see any of this. Silence for a moment as he thinks what to do. Then the rattle of the door handle. How long would it be before he broke the car lock? Or the window? And grabbed my arm? (Strange that I'm able to imagine a black man, not a white one, doing this.)

I decide that huddling in fright in Camelo for the whole night, listening for hostile sounds, is too grim to bear. Better to face the danger immediately. Better to do anything than imagine the worst thing possible and then wait for it to happen. Anyway, I rationalize, Mr. Matthew will be waiting for me. He'll worry if I don't arrive. I decide to walk the rest of the way to Fleur de Lys, which can only be a few miles at best. I pack my pyjamas in my knapsack, put my wallet in my pocket, lock Camelo, and set out on foot.

The night is so dark that I can only just make out the direction of the road. I don't have a flashlight, so I have no idea where my feet are stepping. Every unevenness underfoot is a snake, every rustle in the bush a leopard crouched ready to spring, every unusual noise a man ready to grab me. My breathing is fast, my throat dry. When I feel my way down a small hill to cross a bridge over a stream, the air around me suddenly becomes heavy and moist.

After stumbling along for about an hour, almost numb with fear, a car approaches and I flag it down with infinite relief. No car was ever greeted more joyfully, even though I have no idea who is in it. Luckily the driver is Farnie, the secretary of Fleur de Lys, who recognizes me immediately as the Canadian student. He and his wife have been visiting a veld fire to make sure it isn't a threat to the farm. As we talk, a station wagon drives up and stops. It's

Mr. Matthew, just returning from a trip to Nelspruit. He's a large man in khaki, who greets me pleasantly.

"Welcome to Fleur de Lys," he says, shaking my hand. "We didn't expect you 'til the end of the week." I find out later that he had been extremely worried when he passed Camelo with her Grahamstown licence parked at the side of the road. What had happened to the Canadian student?

"Didn't you get my telegram?"

"No," he says. "Telegrams take days to arrive in Africa."

Mr. Matthew drives me back to Camelo, where I collect my suitcase. "It would probably be broken into during the night if there's anything in it," he explains, "and I don't want you out at night by yourself again." Then we drive to the long side road leading to Fleur de Lys Farm. His bungalow, set among several other low buildings, is spacious and white, resembling a model new home.

"I hope the giraffe will be up to your expectations," he says kindly as he shows me to his daughters' large room with adjacent bathroom where I will sleep. In my paranoid state from my walk in the dark, I'm pleased to see that the door has a lock. I bolt it and within minutes am in bed, fast asleep.

# 5

## First Days at Fleur de Lys

I awake the next morning to find my room bright with sunshine. In the yard outside my window, a small African boy is raking the sandy ground smooth. (How tidy, I think, but later I'll learn this is done to make snake tracks obvious). Farther off in an open shed, an African man perches on a stool milking, in turn, seven Ayrshire cows; an eighth is trying to gore a large dog that prances on the far side of the fence composed of small gum tree poles stuck upright into the ground.

Mr. Matthew, dressed in khaki cotton shirt and trousers, stands beside one of the trucks consulting with the African driver. Seeing him for the first time in daylight, I realize what a commanding figure he is, large in girth and well over six feet tall, his hair white but his face youthful. I can hear him shouting instructions in a language I don't understand. When he's finished, he turns on his heel and strides back to the nearby two-room thatch hut office of Fleur de Lys Citrus Estates. The truck backs up, turns around, and roars off.

I decide to inaugurate the new khaki shirt and long pants I bought in Grahamstown (both made for men because such garb isn't considered suitable for women); the long pants are meant to protect me from snakes, but they're hot—it's already warm and by noon it will be ninety-five degrees in the shade. What a difference in temperature a thousand miles makes!

I unlock my bedroom door and venture into the living room, where a native man is polishing the floor on his hands and knees while another sets the table for breakfast in the adjoining dining room. Except for copies of *Time* and *The Saturday Evening Post* left lying carelessly beside one of the armchairs, the room is immaculate.

"Good morning," I say to the African.

He looks up, startled. "Jaa miss," he mumbles, continuing his work. This is Nelson, the cook's helper.

The view from the front window is beautiful. Far beyond the garden to the west, the bushveld slopes slowly upward to the foot of the Drakensburg mountain range, which rises sheerly and majestically four thousand feet. The rising sun shines obliquely on the crags, etching the grey folds and ridges with black shadows and casting a reddish glow over the whole scene.

While I'm at the window, Mr. Matthew and a young white man, also dressed in khaki, come in to breakfast.

Mr. Matthew introduces him. "This is George Winner (pronounced Vinner), who's in charge of citrus production. He eats with us and boards with one of the married managers."

"I'm Anne from Canada," I introduce myself, nodding to him since he doesn't seem to be into shaking hands. He's about six feet tall and good looking, with blond hair.

We eat pawpaw, cold cereal, fish, and toast served by Watch, the cook, whom Mr. Matthew doesn't introduce. Watch has a shrivelled left arm, damaged since birth, but he manages his job admirably.

"What are your plans?" Mr. Matthew asks me.

"To study the giraffe," I reply promptly. George is hunched over his food but looks up now and then to regard me curiously.

"What aspect of the giraffe?" Mr. Matthew asks.

"Everything I can think of. All aspects."

"How long will it take?"

"Well, that depends on how it works out, I guess," I answer hesitantly.

"I think you should stay three or four months," he says. "It'll take a while for you to find your bearings. We'll do all we can to help you. But why on earth giraffe?"

"I've always wanted to study them, ever since I was two."

George stops eating for a minute. "You really came all the way here from Canada just to study giraffe?" he asks with a German accent.

I nod. He shakes his head in bemusement and turns back to the fish.

"As far as your room and board goes," Mr. Matthew says, "you can pay for them by doing some typing for me. As well, you can make a census of the grasses and another of the game animals on the farm. I've always wanted that done. We've got giraffe, sable, impala, wildebeest, zebra, kudu, sessaby...lots of game and lots of room. The citrus trees take up about sixty acres, and most of the rest of the twenty thousand acres are for cattle—and, of course, for the wild animals which were here before we put in the fences."

I start to thank him again but he interrupts me.

"You can use my field glasses," he continues, "and I'd like you to make a colour movie of the giraffe. I have a 16 mm camera with four lenses you can use." I'm astounded by his kindness. Filming giraffe will be a fantastic addition to my work; I hadn't considered it even possible because I don't have money for a movie camera or for film. I sit back in delighted shock. Mr. Matthew stops me before I can thank him adequately all over again.

"Don't thank me. I only hope that someone would do as much for my daughters if he had the opportunity."

After breakfast, when the screen door slams after George as he departs for the orange nursery, Mr. Matthew guides me into the living room where there are studio photographs of his wife and three daughters, all of whom are overseas pursuing various goals and leaving Mr. Matthew alone. He tells me about each of his daughters at some length, although I quickly mix up their photographs, their names, and what they are now doing. Later, I wonder if he was trying to impress on me that he is a family man—someone devoted only to his wife and children. That certainly suits me. He cautions me to be on the alert for any movement on the ground which might be a snake, and tells me to watch for ticks.

"Everyone gets sick sooner or later, so you will too, but the diseases aren't fatal around here."

After he has closed all the curtains in the house to keep out the sun and heat, Mr. Matthew drives me in his station wagon to Camelo, who sits untouched by the side of the road. When I start her motor she springs into life immediately.

"It must have had an air bubble in the petrol line," he says matter-of-factly. He seems to sense I'm afraid it was something stupid I did that made her break down. Camelo and I follow Mr. Matthew back to the farm, where he instructs two maintenance men to oil and clean her. This is the life!

Before he returns to work, Mr. Matthew drives me on a tour of the farmed area. The orange, grapefruit, and lemon trees (which have no sign of fruit now that the picking season is over) are arranged in a number of discrete orchards separated by windbreaks of poplar and eucalyptus gum trees, the latter used eventually for building poles because of their size and strength. Among the orchards are acres of mixed crops—lychees, bananas, pawpaws, corn, guavas, and vegetables, all tended by African workers who leap to attention when we drive up and answer Mr. Matthew with what seems to amount to a "Yes Sir" or "No Sir" when he asks for details.

"We sell about twenty thousand boxes of oranges a year, most of them to Britain," Mr. Matthew tells me, "and the profits are so good that all sorts of citrus farms are starting up in the lowveld. I'm an advisor to several in Swaziland."

"Won't they be competition for Fleur de Lys?" I ask.

"In time," he replies, "which is why I'm building up the cattle herds. I'm experimenting by fertilizing one of the paddocks to see if this makes enough difference in the vegetation to make it economic. No one has done that before."

"So the cattle will sell for more money?"

"Yes. I don't have time to take you to the dam at the back of the farm where you'll have to be careful. We've had lions, cheetah, buffalo, and even elephants there, not all at once of course, so don't get out of your car."

I spend the rest of the morning strolling around the farm, nodding in a friendly manner to the Africans who are cleaning vehicles, checking fences, or working among the various crops. Most of them return my greeting. I don't go into the native living quarters, where I can see scores of huts among which a few women sweep or lounge, their children playing at their feet. Many of the Fleur de Lys workers live here, but others live in the African reserve bordering Fleur de Lys on the south; all belong to the Shangaan tribe.

Farnie, the secretary who met me on the road the night before, sees me from his window smiling and saluting various natives. He comes out from the office to speak to me shortly after Mr. Matthew has left to drive the short distance to Klaserie to collect the mail. He's dressed in the short-sleeved khaki uniform that all the white men wear.

"You shouldn't make friends with the natives," he tells me sternly. "It won't get you anywhere. You can't trust them. Natives will always let you down—if not next week, then next year or in five years. I've lived in South Africa all my life, so I know."

"I was just saying hello," I reply.

"They'll laugh at you behind your back if you go on like that. If you try and be friends or do anything for them, they think you're a fool."

"No…" I begin to defend them, but he cuts me off.

"They're all filthy, too, so don't get near them; even if they wash, they only wash the parts of them that show. They're clever in a sly way."

"That's because they don't have baths and showers like we do," I say hotly.

"You'll get over your idealism," he insists. "The only thing natives understand is the whip."

"That's disgusting," I say in revulsion.

"No it's not," replies Farnie. "One native whipped his son hard for doing something bad. He said that he was whipped that way by white men when he did wrong, and it was a very good thing."

Farnie, the Fleur de Lys secretary.

George, the Fleur de Lys citrus fruit expert.

Mr. Alexander Matthew, the manager of Fleur de Lys, with his dog, Jessie.

"That's awful!"

"Remember Seretse Khama, that son of a chief from Botswana who studied in England and married an English girl called Ruth? That marriage hasn't worked at all. Of course not, but Ruth's too proud to ask for a separation or a divorce," he states.

"How on earth would you know that?" I ask. "Anyway, it isn't true."

Before he can answer we see Mr. Matthew's car coming back up the lane, so Farnie retreats into the office.

By noon, when the three of us from breakfast gather again for a lunch of salad, rice, and meat that Watch has prepared, it's hot.

"Why don't you have a nap to catch up on your sleep," Mr. Matthew says to me. "After tea I'll drive you out to see some giraffe and to scout for fires."

Indeed, fires will be a constant worry during my early months at Fleur de Lys. It's still winter, the dry season here, with the grass and bushes on the veld like tinder. Every farm but Fleur de Lys has had at least one major fire, and one farm has been completely destroyed. Mr. Matthew has organized firebreaks—wide strips of land with no vegetation on them—all the way around the farm and between the many paddocks or fields. At the first sign of fire on a neighbour's farm, he sends out fifty or so of his "boys" to build a back fire which the neighbour's fire can't jump over.

My first giraffe is a female approaching a small pool of water near the road for a drink. Beside her are four impala, as if present to give perspective to her immense size; if he wanted, a male impala could easily walk right under the giraffe without his horns touching her stomach. (Even I could walk right under her!) The giraffe merely glances at the station wagon as we park opposite the pool.

"She's gorgeous and she's huge!" I whisper, filled with joy to be this close to a giraffe in the wild.

Mr. Matthew smiles and nods. "Wild animals don't connect vehicles with danger or with people," he says.

The giraffe stands beside the pool for a minute, then bends her knees forward to reach the water with her mouth. She sucks in water for a few seconds, then swings her head up quickly to make sure all is well.

We watch the giraffe until she ambles off into the bush, then drive along the road toward a cloud of smoke in the distance.

"I thought she'd put her head up and down slowly," I say to Mr. Matthew. "There's a terrific change in blood pressure when she moves her head down twelve feet and then up again. I wonder why she doesn't faint from the change in blood pressure at her brain?"

Mr. Matthew shrugs his shoulders and laughs. "I have no idea," he admits.

We come to a company truck, where twenty workers are milling around. Mr. Matthew immediately begins organizing them through the two black overseers while I wait in the station wagon. A grass fire with flames leaping ten feet high crackles along beside the road. To my horror, three giraffe are crashing about in the bush ahead of the blaze; they wait until they are almost burned then dash to the west, where their heads are silhouetted against the dull red setting sun. Once beyond the fire, they stride calmly on.

When the wind changes direction so that it's blowing away from Fleur de Lys property, Mr. Matthew leaves several men on watch while the rest of us return to the farm buildings. To have seen so many wild giraffe on one afternoon makes me euphoric!

The next morning, after breakfast, I begin my giraffe work in earnest. I have permission to drive along the dirt tracks through all nine paddocks where giraffe live, or even onto the veld itself if there are few trees or bushes. Each huge enclosure has a water supply where cattle can drink, often in the form of a borehole and trough at the junction of four paddocks. Mr. Matthew tells me that one such junction is called "Giraffe" because these animals often come there for water, so I decide to make this my first destination. It's about four miles from the farmhouse. On the track through Paddock 7, I stop Camelo to watch a male giraffe about twenty yards away. He's so huge that, with Mr. Matthew's field glasses, I can only examine sections of his body at any one time. Before long I notice five more giraffe, all camouflaged, all browsing on trees around me. I am amazed that I didn't at first notice such large animals, but if I'm driving along a track my eyes are focused on the road's bumps and holes; when I do glimpse a giraffe's leg out of a side window, I must take it for a small tree trunk. One of the giraffe stares at Camelo but the others ignore her. Finally, when they wander away from the track, I continue on to the borehole.

As I approach the water trough, I disturb a giraffe who canters a short distance away; I park under a tree and settle down to see if this male will return. He watches us seriously for a while, then edges closer and closer to the trough. Finally, in a fit of daring, he spreads his front legs slightly and leans down to drink. After a few sips he lifts his head quickly to study us for another five minutes. Then he drinks again at more length. Giraffe watching is going to take lots of patience!

When he has drunk enough, the male strides toward the bush and is just disappearing from sight when another big bull arrives. He ignores the water, instead jumping gracefully over two four-foot barbed wire fences to follow the first male. A friend? An enemy? There's no way to know. But surely, eventually, when I've spent enough time watching them, I'll be able to tell what's on their minds! I write down what I've seen in my notebook and judge my morning's work to be a great success, as does Mr. Matthew when I see him at lunch. I've observed a number of giraffe browsing and I've watched one drinking from a trough where he didn't have to lean down very far. I know where to find giraffe and how not to scare them away.

In the afternoon I drive through several paddocks without seeing any giraffe, then decide to go back to the waterhole where I saw the female drinking earlier, and where all animals have to bend right down to reach the water. I want to watch a giraffe's exact movements without being hypnotized by his or her very presence as I had been the day before. There we sit for three hours, Camelo and I, baking in direct sunlight. A family of monkeys, a mongoose, four male impala, and a warthog, each nervous and alert, come to drink before dark. But no giraffe.

The ninety-five giraffe of Fleur de Lys almost all live in the eastern section of the property; the half west of the main road is too heavily farmed and, in addition, has many fences, an inhibiting railway line and, as I'm told later, an aggressive black mamba. Many giraffe frequent the eight paddocks where the cattle are mostly kept, but the ninth, most southern paddock, called the Game Reserve on the south side of the Klaserie River and the Guernsey Area on the north side, is far larger than the others—about four miles long by two miles wide.

"The Game Reserve and Guernsey Area teem with game," Mr. Matthew tells me at dinner. "There are lots of giraffe there, as well as the herd of sable that poachers have been after."

"Do you keep cattle there?"

"Not usually," he replies. "I want to keep the whole paddock for game, and make this the focus for a private game reserve where no animals may be killed."

"That would be great!" I exclaim.

"Before farms put up fences around here during the past ten years, giraffe wandered over long distances, but now a group stays in the Game Reserve area for a year or more. Or forever, perhaps, since there's lots of food and water from the Klaserie River." I make a mental note to check out this area for giraffe next, as it seems likely to yield fruitful results. It's more open than the other paddocks, so it will be easier to watch a number of giraffe and what they're each up to at any one time. As well, the Game Reserve has a circular road that makes it easy to drive in and out.

"When we fenced one of the other paddocks a couple of years ago," Mr. Matthew continues, "a herd of wildebeest [large black antelope with crooked horns] were trapped. There was a trough for watering the cattle in the paddock so we weren't worried about them, at least until we realized that the wildebeest were afraid to drink there. We took down the fence on one side of the paddock so the wildebeest could escape, but they'd got used to it and stampeded back into the paddock every time we tried to drive them to freedom. We finally had to shoot them, which was too bad. The natives had lots of meat, though, and we made fly switches from their tails."

The next morning at dawn, I drive into the Game Reserve and park under a tree. Even before I've turned off Camelo's motor I see five giraffe about fifty yards away—two adult males, two smaller females, and a half-grown male, all browsing in the vicinity of a large fig tree which I will later identify as *Ficus petersii*. None pays any attention to me. The animals snatch a few leaves from one bush or tree, then move on to another as if they're supremely fussy about their diet. Presumably their selectivity evolved so that no one plant is completely denuded of leaves and therefore unable to survive. Or is the best diet one with as many different species of leaves, and therefore nutrients, as possible? With my binoculars I scrutinize a female sampling the foliage of a thorn bush which has hundreds of long thorns, chewing each mouthful slowly and carefully to avoid puncturing her mouth. I'm actually watching a wild giraffe, I muse happily. Now, how to study them?

I take out my notebook, draw a rough map of my surroundings, then mark in the individuals with "X"s (males) and "O"s (females), the symbols preceded if necessary by a "y" for young and a "b" for baby. Rough, but it's a start. I can tell the "X"s from the "O"s not only by the presence or lack of a furry penis sheath, but by the hair on their horns. All giraffe have black hair circling their horns when they are born, but the males soon wear off this hair during their frequent head-hitting sparring matches. Adult male horns are completely bald at the top, while those of females, who never spar, retain their hairy crown. (When they are born, the young have horns of cartilage which, to facilitate birth, lie flat under the skin. Within a day or two they stand upright on the skull, and later on they gradually turn into bone.)

When I finish my diagram I survey the area again. Now the larger male and a female with jaunty horn hairs are browsing nearer me on the small leaflets of an acacia tree. The male has body spots with jagged edges rather like our symbolization of stars, so I call him Star. His companion becomes Pom-Pom, named for her prominent horn hairs. The other three giraffe have drifted toward the north, where the young animal is nibbling at a thorn bush, and two males are standing side by side, head to tail, hitting each other lightly with their horns. I note the time, thirty minutes since I arrived, and mark down these new activities. Soon I will set up my notebook so that I can note down at five minute intervals throughout the day exactly what each animal I can see is doing.

Giraffe, because of their size, spend most of their time eating, which is boring to watch, so I'm pleased to realize that activities such as sparring are fairly common. When I turn my attention again to the pair of males, one is drawing his head along the body of the other—getting rid of ticks? a gentle assault? His opponent then reciprocates the activity. No real aggression here. After a pause, the smaller male gives the larger a sharp blow to his front leg, knocking it out from under him. The larger retaliates with a swat, horns foremost, at the other's neck. They hit each other more strongly for several minutes, although some strikes miss entirely if the intended receiver shifts his body away, before the sparring match again subsides. The two males stand still for a moment, then wander over to browse at thorn bushes near a lone female.

(Ideally, had I longer to study these giraffe, I could have identified each of them by their spots, as my colleague Bristol Foster was to do in the mid-1960s in Nairobi National Park; during a three-year study, he toured the park weekly to note both the location and the associations of the animals. He photographed each giraffe from his or her left side, eventually accumulating 241 photos which he took with him into the park to identify any giraffe he encountered. I later used this technique on a study of individual interactions among eighteen giraffe in the Taronga Zoo in Sydney, Australia. I knew that their body spots didn't change shape as an animal grew up because I found pictures of most of them as babies in the local newspaper.)

Later that week I decide to stroll away from Camelo to see how close I can come on foot to giraffe before they flee. I saunter silently toward the two males who are sparring in the distance, waving their necks about as first one, then the other slams his opponent with his horns. They are so preoccupied that I can settle down under a bush to stare at them through field glasses without being noticed. I watch them for half an hour, noting down at five minute intervals exactly what is happening between them and what the other giraffe beyond them are doing—all browsing, of course, because in this dry season sufficient food is hard to come by.

After sparring for a while the males start feeding too. They are joined by seven more giraffe, each sampling leaves from one tree or bush, then another, nibbling their way slowly toward invisible me. Three wander within forty yards of me while I remain motionless, looking as much like a bush as possible. At the same time seven kudu cows and a kudu bull approach from the right, munching away on the grass; they are so close I can see their eyelashes. I decide to make this a study of communication. When one giraffe or kudu sees me, how will he or she alert the others?

This heaven of being unseen among wild animals can't last. Eventually the first giraffe to notice me freezes, staring at me without blinking. She doesn't move suddenly or make a noise to alert the others, but gradually they become aware of her rapt attention, look where she is looking, and begin to stare at me too. Occasionally one steps closer to see better. I feel self-conscious, but keep perfectly still. One of the giraffe snorts to see what this apparition will do. (Ha, ha. I've read that giraffe never make any vocal noises but now I know that's wrong. Later I hear them grunt, low, and wheeze as well.) I remain immobile. Finally, the tension is so great that they all wheel as one and canter off, several turning back a few minutes later to continue staring.

About the same time the kudu also see me, but I've been too preoccupied with the giraffe to notice if or how they communicated my presence to each other. Maybe they saw the giraffe gazing at me and followed suit. They stand in a semicircle with their ears out, wondering what on earth I am. One female barks at me, making little rushes forward. Then they too turn away. Later that day I stalk thirteen wildebeest who are much shyer; unlike the other species, they and zebra are sometimes shot (and eaten) because they may spread disease among the cattle when they drink at the Klaserie River or at the dam. On the walk back to Camelo I meet six zebra, but they too wheel away in disarray. All of the animals are dismayed to encounter me, but I've never been so happy in my life.

Over the next days and months I adopt a heavenly, if arduous, routine. Mr. Matthew knocks on my door each morning at 5:30 with a cup of tea, and by six o'clock I'm driving Camelo to the Game Reserve to see if Star or Pom-Pom are about, or motoring, at least once a week, through the various paddocks hunting for other giraffe whom I can watch during the day. By the time I've found some, it's usually near breakfast time at eight o'clock so I return to the farm to eat and to report my giraffe sightings to Mr. Matthew. I then spend four hours in the field with the giraffe before lunch, and another four hours in the afternoon, until it's dark. I'm soon able to recognize a number of individuals besides Star and Pom-Pom such as Lumpy, who has lumps on his lower neck; Cream, whose body spots rest on a cream rather than a white background; and Limpy, who has a snare embedded in his lower leg.

When I have a few spare minutes, I read newspapers to see what apartheid supporters are up to and write letters home which I can finish, if necessary, while watching giraffe stand about in a kind of stupor and snooze in the middle of the day. I write each week to Ian, although I'm annoyed at him because, after a few letters in which he ruminates about my suggestion that he come to Africa for a visit or to teach, I hear nothing from him for a month. Even though he wishes me "love" when signing his name, I write crossly that if he can't manage to send me a letter once a week, then surely there won't be much of a future for us. (After all, romantic couples in biographies often correspond with each other every day, if not twice a day, thanks to a marvellous postal service.)

I also write to several possible contacts in East Africa to see if I can visit the giraffe there after my four months at Fleur de Lys are up. I wonder if I might work out of Arusha, a small town in Tanganyika (later Tanzania) and headquarters for the Serengeti National Park, where giraffe abound. Giraffe are as numerous in East Africa as anywhere on the continent; from Arusha I'd be able to compare the habitat, the giraffe, and their spotting with those at Fleur de Lys.

Each evening at seven o'clock, after I shower and change into a blouse and one of my three cotton skirts, Watch serves dinner to Mr. Matthew, George, and me. Jessie, Mr. Matthew's rotund brown dog, sits beside Mr. Matthew's leg, hoping for a hand-down of food. After dessert George heads for home, and although I'm exhausted, Mr. Matthew and I retire to the living room to chat while Jessie settles beside my stuffed chair to be patted. At first I stroke the fur on her back with my hand and then, when my arm is tired, with my foot. Before long she has twisted about so my foot is rubbing back and forth against her teats, which embarrasses me, so I stop my attentions.

Mr. Matthew meanwhile talks steadily until nine o'clock when, after I've secretly consulted my watch, I feel free to say I'm tired and go to bed. He tells me about his love for Scotland, his ancestral country, where he attended university at Edinburgh; his wife, whom he met and married while doing graduate work in agriculture in Berkeley, California; his three daughters, who are in Britain and in America, along with other relatives; his take on the agricultural scene in the Transvaal; his adventures with wild and domestic animals; and his often negative dealings with natives.

"When the Africans aren't allowed much schooling, how can you expect them to be smart?" I ask, when he has commented on how dense most of them are.

"They don't need schooling for the jobs they do. Europeans earn more money but they work far harder," says Mr. Matthew who himself starts each day at 4:30. "Most boys don't take responsibility, so we have to keep checking up on them. They'll forget to lock a shed or to check the air pressure in car tires."

"But if they had schooling they'd be more responsible. They could get better jobs and make more money," I protest.

"No," he insists. "You can send a boy to Oxford, but he's still from the Stone Age and just as hopeless as the other boys who haven't been educated."

I register my disbelief with a shake of the head. Mr. Matthew is talking the same nonsense as the policemen were on the ship. How can millions of white people be brainwashed to accept the same set of crazy ideas? The whites all have at least some education, so why can't they see that their beliefs are only valid because it's to their great advantage that they be accepted as truth?

"They aren't like us," he explains carefully, as if to a child. "Don't worry about them. Our workers and their families have everything they need at Fleur de Lys and are happy enough. They just aren't capable of learning much."

"But we don't know that. In the United States many blacks have responsible jobs and do good work," I insist.

"And millions more have poor jobs or no jobs at all. I know. I lived there for years."

I decide not to mention natives to Mr. Matthew in the future. How can a man so thoughtful and kind toward me be incapable of seeing that natives, too, need a chance?

Mr. Matthew has worked with Africans all his life, but he is dead wrong about their abilities, as Drury Pifer reports in his book, *Innocents in Africa*, about his father, an American from Seattle who worked with native labourers in the Transvaal gold mines in the 1930s. At first, Mr. Pifer decided that most of the men lacked "instinct for even simple machine work." He saw some who struggled with bolts and nuts for months, yet were still unable to put the nut in the bolt and turn it in the right direction. But after the men came to trust him because he was an American who treated them as human beings, the problems with nuts and bolts evaporated. He was soon running the most efficient and productive crews in the mine.

Before I'm a week at Fleur de Lys, I see my first snake. I've talked about snakes with Farnie, the secretary, who has moulted skins from mambas, cobras, and puff adders hanging in the office. It's a subject that obsesses me because I'm terrified of being bitten by one in the veld and dying before I can reach help.

"There are snakes about, but your chances of seeing them are slim. I bet you'll only see one the whole time you're here. Snakes are more afraid of men than we are of them," he says.

I try to believe this, though I'm not a man. "I guess I'm afraid of them because they're so deadly. There are no poisonous snakes where I live in Canada."

"Even if you're bitten, we keep an anti-venom kit in the refrigerator," Farnie says.

"But I've read that mamba poison can kill a person in a few minutes. And they can move faster than a person can run. If I'm bitten while watching giraffe, I won't have time to get back here."

"That's true."

In the ominous silence that follows his remark, I glance down at the documents on the desk.

"In this census sheet on cattle deaths there's a column for snakebite," I say, pointing to one of the open pages of a binder. "And look, there's a tick in that column!"

"There are certainly snakes about," Farnie concedes.

"When I was little I used to get books on snakes out of the library," I tell him. "I used to make sure the pictures of rattlesnakes weren't open before I went to bed, so that none of them would escape from the book during the night."

The next afternoon as I'm getting Camelo ready to drive to the giraffe, there's a shout. One of the young lads has seen a spitting cobra, called a *rinkhals* in Afrikaans, crawl under Mr. Matthew's station wagon. I retreat to the wall of the house, twenty feet away from his car, to watch. Surely I'm safe when there are other people between me and it?

"Bring my gun," Mr. Matthew calls to George.

"Drive the car out of the way," he orders one of the natives. As the station wagon jerks forward, its left rear wheel runs over the four-foot long snake. It writhes quickly and then more slowly.

"That got it," George cries, handing Mr. Matthew the weapon.

"No, it's still moving," exclaims Mr. Matthew. But as he steps forward and blasts the snake with his shotgun, Jessie leaps forward, too. The loud bang is followed by her yelp.

"Get the dog out of here," Mr. Matthew shouts, too late. The snake is dead but Jessie hunkers down in the dirt with her head between her paws, whimpering.

"Maybe the snake got her in the face," George says, kneeling beside the dog. Spitting cobras protect themselves by projecting venom at the eyes of threatening animals, including people.

"She must have been worried about her puppies," Mr. Matthew says, patting Jessie. "Anne, get a wet cloth and some boracic acid, and we'll wash out her eyes," he tells me.

He wipes Jessie's face and eyes carefully, deciding that it was only sand in them, blown there by the shotgun blast. I'm breathless with the excitement, but everyone else is blasé. Soon they're busy again about their own affairs.

"Wow," I say to Mr. Matthew who is putting away his gun. "I'm glad the boy saw the cobra."

"He'll get a bonus," he says. He leans down to give Jessie another pat. "The last time you had pups, it was a puff adder we had to kill," he teases her, ruffling the fur between her ears.

"I guess if I'm only going to see one snake while I'm here, that was it," I say to Farnie in mock relief. When on my field work I'll have to choose between tramping noisily to scare away snakes and creeping quietly so as not to frighten giraffe. I have a feeling I'll be moving less quietly than I did my first week.

# 6

## Settling in at Fleur de Lys

Surely the places that attract giraffe are important? Giraffe occur in some paddocks but not others, and congregate in preferred areas of these paddocks. Why? I decide that my method of study should include not only documenting what giraffe are doing at five minute intervals each day, but where they are doing it. To this end, over the course of a week, I make a map of the area at Fleur de Lys where giraffe hang out, mostly by walking through the veld counting my footsteps (which equal 0.81 yards from one step to the next) and plotting the angles of tracks and fences in the various camps. On this map I give the areas within each paddock a letter from "a" to "1," with "a" in the middle and the other letters anti-clockwise from the upper right-hand corner.

At the same time, I map the various types of vegetation in the paddocks where giraffe occur but also where they don't; the plants in each area will probably explain, at least in part, giraffe distribution. To do this, I stake out an area of ground, chosen at random in a paddock as I'd been taught to do at university, then count all the plants (except grasses, which giraffe don't eat) that grow there.

All the paddocks, with the exception of the Game Reserve, are covered with what Mr. Matthew calls bushveld, which is made up of mostly thorny bushes growing above the grass (I calculate an average of one thorn bush for every forty-two square yards), but with some trees as well. The trees aren't all the same (mostly belonging to six species), nor are the bushes, of which there are at least ten species. Since the bushes are at eye level and far more voluminous there than the tree trunks, the word "bushveld" perfectly describes what exists. The most common bush (which I later have identified) is *Acacia swazica,* fancied by giraffe despite the thorns jutting out among the leaves.

The Game Reserve, by contrast, has four different types of vegetation—open area, thorn bush, parkland, and bushveld. The open areas now covered with grass are a legacy of farming by natives until 1950, which had caused heavy erosion. These areas have no trees at all, which makes giraffe easy to see, but at the same time makes it rather pointless for them to loiter there. They are bordered in some places by dense clumps of thorn bushes of several species—a bush for every three square yards—all less than ten feet tall. The giraffe spend much more time here because they like thorn leaves, even though they have to bend low to reach them and be wary of the thorns, which presumably evolved to curtail browsing by animals.

(While examining this area I notice various holes in the ground, each made by some animal, which solve my problem of soiled menstrual napkins. Each month I stuff these down different holes; I don't know what else to do with them because I live in an entirely male household where Nelson regularly goes through the garbage to salvage anything useful. I do this even though I feel badly to be so uncivil to the resident animals, and especially don't want to annoy mamba or cobra occupants.)

The giraffe in the Game Reserve largely spend their days in the other two types of vegetation, bushveld and parkland. The parkland where I first saw Star and Pom-Pom is dotted primarily with trees; within a circular area with a radius of seventy-five paces from a base marula tree which I stake out, I count thirty-eight trees or bushes. The density of trees alone is about one tree for every thousand square yards.

I'm already documenting giraffe behaviour by noting the activity of each animal at five-minute intervals. This will allow me to tabulate how much time they spend browsing, chewing their cud, lying down, walking about, and interacting with other individuals. Eventually I'll be able to tabulate differences among the activities of young, males, females, solitary individuals, and giraffe in groups. Once, when the giraffe are half a mile away and I'm hot and restive between counts, I climb out of Camelo to practise ballet moves using the door handle of the car for a bar. Surely there's no harm in this? But there is. A female, immediately curious as to what I'm doing, approaches slowly, watching me steadily all the while. She comes within forty yards of Camelo without stopping. Because I have no category for "curiosity" or "staring," I reluctantly climb back into the car.

I decide also to collect samples of leaves the giraffe eat so that I can later have the species identified. This is an ideal time for a broad survey of browsed species because, in this dry season, animals have a hard time finding enough

A browsing Fleur de Lys giraffe.

food. The favourite trees of giraffe have already been stripped of leaves to a
height of fourteen feet or more, so they're forced to munch on less-favoured
plants. Such a survey is tricky to carry out: I have to be careful to identify
with certainty exactly the tree or bush a giraffe is sampling perhaps twenty yards
away, even though the animal rushes off in a panic when I leave Camelo to
mark the plant.

I buy lengths of yellow and red cotton cloth from the tiny store at Klaserie,
the three-hut "town" just outside Fleur de Lys. (No wonder it isn't on any
map!) The Indian merchant there makes a bare living selling a small assortment of tinned foods, farm implements, cloth and clothes, and odds and
ends.

"Look," I show Mr. Matthew in the yard after lunch. "I'm going to cut
this cloth into strips to tie around trees and bushes that giraffe and cattle use
for food. Isn't that a good idea? Yellow for giraffe and red for cattle."

"You'd better make the strips thin so the natives won't think of some other use for them and steal them," he says.

I ignore this negative comment. "Before I leave Fleur de Lys, I'll collect fresh leaves from the plants with ribbons around them and have them identified. Maybe at Rhodes University." I think of Dr. Pocock the botanist. "I can collect leaves from non-eaten plants, too, so that we'll know what giraffe *don't* like."

"Good idea. I'll get some of the rangers to help you."

Mr. Matthew calls to a native man who comes over to us. "This is Mokkies, the head ranger," he says to me. "I'll ask him what he thinks."

He turns to Mokkies, whom I've noticed riding a bicycle through the paddocks, a short man in the khaki uniform of a ranger with a red, woolly hat on his head despite the heat. Mr. Matthew explains to him what I want in Fanagalo or Kitchen Kaffir, the vernacular spoken by everyone but me on the farm. Mokkies looks bemused, but vaguely interested. Maybe it will make his day less boring?

"I've told him you're working for the government to save a lot of explanation and keep things simple," Mr. Matthew says. When I think back on his remark today, I'm perplexed. Surely he was aware that the government was no friend of the Africans. There had already been mass protest meetings of Africans and massive school boycotts because of the Bantu Education Act, which restricted their schooling. Already the South African police had gathered enough evidence to soon arrest hundreds of Africans and some whites on charges of treason against government policies. How could he not know this? Or did he believe that his workers were too simple or too far removed from urban centres of revolt to be aware of such black protest? Indeed *were* his workers aware of what was going on in the rest of their country?

At the time I just say "Oh" in response to his comment, wondering if it would not be better to be unconnected with the racist government, but unwilling to say this to Mr. Matthew. I figure he must know what he is doing.

"Mokkies will have some of his boys help him notice what plants the animals are browsing," Mr. Matthew continues.

This all seems so easy that I decide to ask Mokkies and his men to also note down the giraffe they see and where they see them as they ride their bikes around the property—they tour every paddock daily to check for signs of poachers, or broken fences, or other problems. Then I can quantify what kinds of giraffe associate with others (sex, size of individuals, size of group) and where giraffe are likely to be found (perhaps near their beloved thorn bushes, or by water, or in cleared areas where they can see predators). I try to explain this in pantomime to Mokkies, but he only laughs; he has no idea what my gesturing means. I trust that Mr. Matthew will explain later what I want.

I buy a package of four-by-six-inch filing cards at Klaserie and mark off columns so that the size and sex of individual giraffe they see together can be noted. Then I print out the local names for these terms using a Fanagalo dictionary ("man" giraffe and "woman" giraffe). When I show the cards to Mr. Matthew, he bursts out laughing.

"Mokkies can't read," he chuckles. "None of the boys can." He sees this as a simple statement of fact, not an indictment of apartheid.

I take more cards and draw pictures of male and female giraffe, each in three different sizes. I don't have to emphasize their genitalia, thank goodness, because they can be sexed also by their horns—bald for males and hairy for females. (Will they think I'm mad to emphasize the horns for identifying sex and not the genitalia?) While I'm at it, in the corner of each card I make a symbol of the sun and a curve representing its position in the sky so that they can indicate the time of day the giraffe were seen. I can't figure out how to signify date or place, since Mokkies has no idea how to read a map.

"It's getting rather complicated," Mr. Matthew comments mildly when I show him my latest handiwork. "I don't really think cards will work. Mokkies and the others live in dirt huts where they're bound to get crumpled and torn."

He thinks for a moment. "Why not use plaques of wood? We can drill a hole at one end for a thong and Mokkies and the boys can wear them around their necks as they walk or bike around. They can draw pictures of what kinds of giraffe they see together."

"That would be great," I enthuse, annoyed that I had wasted so much time fiddling with cards.

"I'll ask the boy in the workshop to sandpaper some plaques."

I drive out to giraffe-watch, imagining with delight the huge amount of information I'll collect from my new helpers.

One morning visit to the Game Reserve, I take with me Mr. Matthew's movie camera complete with colour film, film I've never used before because of its expense—£4 for one hundred feet of the 16 mm. As if they realize the importance of the occasion, the giraffe respond with a magnificent cornucopia of behaviours. Beside a large marula tree, I find three big males hanging out. I'm able to film Star and another male walking—their two long legs on each side swinging forward almost together so that the hind one does not hit the one ahead. Giraffe don't trot or pace because of their sloping back and long legs, so if they're in a hurry, which they rarely are, they canter or gallop at speeds of up to thirty-five miles an hour. I later film Cream galloping, and later still, when I am back in Canada, analyze walking and galloping gaits from the film. In his walk, Star's neck swings forward twice per stride, coordinated with the swing forward in turn of each front leg. In Cream's gallop he swings his neck and head forward only once during a stride, as his feet leave the ground for a period of suspension; he pulls his neck back to slow his forward momentum when his feet touch the ground again.

I capture Cream browsing at a tree, his head filling the entire frame with his long black tongue reaching up, straining to curl around and grasp a leaf above. When he moves over to a thorn bush, I film him chewing while using both his tongue and his mobile lips, covered with dense fur to protect them from the thorns, to separate leaves on a twig from the thorns. Sometimes he closes his mouth over an entire twig and jerks his head away, stripping the leaves into his mouth. This combing action is aided by the unique extra lobe of the outer tooth of the lower incisor row, which increases the row's width; these incisor teeth meet the upper bony plate which, during evolution, has replaced the upper incisors in most large herbivores. As I'd noticed earlier, the giraffe takes only a few bites from one plant before moving on to the next.

I'm waiting in Camelo with the camera balanced on the window edge to film a female jumping over a fence ahead of me, which she seems on the brink of doing, when I notice two males sparring to the west whom I also want to film. What a predicament! I forget the potential jumper and instead lean out the window to aim at the sparring giraffe, who are waving their necks about between blows—no messy consequences here such as blood or concussions that make human boxing so horrific. Soon I recognize the males as Star and Lumpy. After sparring for a while, both start to browse side by side at a thorn bush.

I turn back to the fence to film the jumping, but the giraffe who may or may not have crossed it is now gone. When I glance back at Star and Lumpy, they are sparring again, but with slow movements lacking intensity; it's as if they're moving underwater. I focus the camera on them and push the button, but their activity is so leisurely that within a minute, although nothing much has happened, I've run out of film. With annoyance I put the camera on the seat beside me and turn again to watch them. Star touches Lumpy's neck gently with his horns, then draws his head lightly down his neck and along his body. Lumpy rubs his cheek back and forth on Star's flank. They rub their necks together slowly, up and down, up and down. Star walks behind Lumpy, pauses for a moment and then, with his pink penis sticking out from its sheath, mounts Lumpy who takes a small step forward at the impact from his rear. After only a few seconds, Star dismounts and he and Lumpy walk off into the bushveld side by side.

I'm amazed by what I've seen. I know virtually nothing about human homosexual behaviour, it being the 1950s, but have assumed from our homophobic Western culture that it's something men do with each other because they are bad. Yet here is my favourite giraffe doing apparently the same thing, and certainly neither Star nor Lumpy is bad. What's going on? I include my observations of homosexual behaviour in giraffe (which is fairly frequent at Fleur de Lys, where there are more males than females) in my report on the giraffe published in 1958. When my aunt reads it, she is aghast. "How could they let a young girl see something like that?" I overhear her telling my uncle. We truly live in a homophobic world. And who are "they"?

(In 1983, anxious to do what I can to combat homophobia which affects and distresses some of my students, I publish the first extensive literature review of homosexual behaviour in wild animals, which describes it in over one hundred species of mammals as well as in many birds and reptiles. It certainly occurs in far more species than that, because some biased zoologists who have observed it have omitted mentioning it in their articles and books, perhaps nervously deciding that it must not really be homosexual behaviour at all. And, of course, the behaviour of only a few of the myriad species of animals in the world has ever been observed in detail.)

As I drive home for lunch, I'm excited at the thought of telling Mr. Matthew and George about the excellent colour footage I've taken of walking giraffe and of giraffe feeding and sparring. But how can I tell them about Star mounting Lumpy? I want to keep my world as asexual as possible, so that the men around me don't think in terms of male and female, and that the place of the latter should be restricted. How can I even describe what I've seen? I've never spoken the word "homosexual" in my life, although I'm a zoologist; I'd be embarrassed to death to describe this and I'm certain (although this is ridiculous) that Mr. Matthew and George would be shocked out of their minds. Except in my giraffe report, for the next twenty years I never mention the homosexual activity I've seen in giraffe to anyone, even my husband. Mr. Matthew must have read about it when I sent him my report on the giraffe in 1958, but he never mentions it to me in his letters.

I've discussed my documentation of the vegetation on Fleur de Lys in detail with Mr. Matthew, and he says he wants me to visit the surrounding countryside to have a better idea of the kind of habitat that is *not* suitable for giraffe. This is what he says, but he may only want company on some of his trips. As a technical advisor to a developing citrus business and irrigation scheme in

Swaziland called "United Plantations," which he inspects periodically, he suggests that I come with him there for a day's visit. Although it's only about two hundred miles south, there is no evidence that giraffe ever lived in Swaziland so its vegetation should be of interest.

Most of the time we drive along in companionable silence. On the rare occasions when a car with its trail of dust approaches, he gives a reflexive jerk on the steering that shifts the car to the left, then rolls up the window to keep out dust while I do the same.

"I wouldn't rest my arm on the window sill," he comments once as I roll down my window after a car has passed. "A man was bitten by a snake that way." From then on I keep my arm at my side.

The view on the way is marvellous; from some of the mountains we cross we can see to our left the whole south of Kruger National Park, and as far beyond as Mozambique. The South African natives don't glance at us as we pass, but when we enter Swaziland the Swazis do, often waving and smiling. This is surely because their country more nearly belongs to them. Swaziland is a protectorate of the British government, ruled over by a British High Commissioner based in Pretoria; it will gain independence in 1968.

While Mr. Matthew talks to the manager about the actual running of the company, the managing director shows me over the plantation in his Jeep; he and his wife have just flown down from Pretoria for a few days to see how things are going—they have houses in both Pretoria and Swaziland. United Plantations involves a huge river diversion scheme which will eventually place forty thousand acres of land under irrigation from an irrigation canal running three and a half miles, and ending at turbines to supply electricity. The company is already growing rice, citrus, and bananas, clearing the amazing sum of £140 per acre from the bananas. Soon it will produce pineapples as well as, on the mountainous part of the estate, pine forests for lumber. This company still flourishes to this day at Pigg's Peak in Swaziland.

As the managing director floods me with facts and figures at the end of our tour, I watch Mr. Matthew perform. Men rush about saying "Yes, Mr. Matthew," and "No, Mr. Matthew," and "I'll get back to you on that, Mr. Matthew," and "If you say so, Mr. Matthew." Mr. Matthew has a master's degree in agriculture from the University of California at Berkeley, so he is a well-respected expert.

On a Saturday, Mr. Matthew takes me over to meet and have dinner with a neighbouring "European" couple who own a store at Acornhoek, five miles away. (In South Africa, "European" is a synonym for "white skinned"; most South African "Europeans" have never been to Europe).

"Acornhoek vegetation looks like that at Fleur de Lys," he explains to me so I'll know this is not a trip to expand my sense of the environment, "but you should meet some of our neighbours." At the time, this didn't seem as unusual as it would have been for him to go to dinner alone, leaving me at home to be served by Watch. Later, I realize that it was tricky for him to appear with a strange young woman, as if he had repudiated his wife which was certainly not the case. Would his friends understand?

"I'm sure you've been missing female chatter," he comments on the drive over.

"Not really," I say. I would have had nothing in common with the immaculately dressed director's wife in Swaziland, whom I had met briefly.

"Mrs. Matthew and I used to have dinner at Acornhoek every week. We arranged a signal for when to escape. I'd say 'I guess we really should be getting back,' and then she'd say 'I'll just finish these few rows of knitting,' as if she hated to tear herself away. It worked pretty well." He chuckles at the memory.

The couple at Acornhoek greet me kindly, treating me as an equal, except that I call them by their formal names, and they call me Anne. The evening isn't exciting, but we do chat a bit about animals and about natives, my two prime interests.

"South Africa would be much better off economically if we'd never used natives for labour," Mr. Matthew says over dessert as the others nod. It doesn't seem the right place to ask what else they could be used for, and why it should be up to some people to find a use for others.

"I know the Nationalists are hard on the natives," he goes on, "but at least the natives know their place. At least the Nationalists are trying to do something positive. When the English liberals were in charge things were much less organized. The natives didn't know where they stood." Again the others nod while I sit silent.

"I saw sixty-two giraffe in one afternoon," the wife says when she hears that this is my species of choice. I note down where and when this happened for my files.

Later, we talk about other game. "Poachers have been after the sable again," Mr. Matthew reports, referring to the twenty-two sable antelope living in our Game Reserve paddock where I'm studying giraffe. They're rare antelope with magnificent, backward-sweeping horns.

"I guess lots of people know you have a herd," says our host sympathetically.

"They've already taken two of the biggest males," Mr. Matthew complains. "The horns would be a good five feet long."

"I'd shoot the poachers if I caught them," snarls our host, who doesn't seem to be joking.

As we leave Acornhoek about 9:30, we're horrified to see a fire in the direction of Fleur de Lys. When we reach our property, we find that two hundred acres of a hill and valley in the Game Reserve have been burned, with flames still spreading in all directions. In the black night it's a spectacular, if horrific, sight. Mr. Matthew is aghast because there's no hope of putting it out, with the nearest fire break nearly a mile away; fifteen hundred acres is at risk. I feel hysterical because the Game Reserve is home to Star and Pom-Pom and their friends. We rush to the farmhouse to find that George has organized a truck and gone to the native compound to root out twelve unhappy "boys" from a drinking party. Just as the truck sets off for the Game Reserve, with Mr. Matthew and me following, it starts to pour rain; when we reach the fire five minutes later, it's effectively out. This is the first rain for a month, so its timing is miraculous.

We stay near the fire for a while to make sure the flames won't flare up again, Mr. Matthew and I in his car, George alone in the truck cab, and the native "boys" crowded under the truck to keep dry. I notice Mokkies among the others and nod to him, but he doesn't give any sign of recognizing me or wanting to report any progress on the giraffe-recording front. Perhaps he doesn't like to be singled out from the others as my helper? As we wait, a rattly vehicle without headlights comes bumping and banging along the rough road at about forty mph, its noise preceding and following it long after it has passed into darkness. How can the driver see the road when it's so dark? If it had hit our truck, the men under it would have been crushed.

"What was that!" George cries out.

"Heaven knows," Mr. Matthew calls back.

Nothing more is heard about the vehicle and the episode is forgotten. Excitements like this would be discussed for days in Canada (That truck was going far too fast! Who could have been the driver? Why wasn't he charged for driving without lights? Was there some illegal connection?), but cause no stir at all in these parts.

"I bet the fire was set on purpose," Mr. Matthew says to George, who has wandered over to our station wagon and is looking in the window.

"I doubt it," he replies. "One of the boys who lives near here was having a beer party—it was probably an accident."

"A few years ago there was a small Cheeza Cheeza movement in South Africa in sympathy with the Mau Mau protest in Kenya," Mr. Matthew tells me. "*Cheeza* means fire, and these gangs went around burning the crops of white farmers."

"That could be disastrous," I exclaim.

"Yes," he agrees, "but they never got very organized, thank goodness. Last year there was a fire over there, half a mile from the road. We had to bash the truck through bush and across a river to get to it. When we had that one out, another started further over and when we got that out, there was a third."

"Did you find out who started them?" I ask.

"No. But we were pretty mad."

In his eagerness to have me see more giraffe habitat, and certainly because he wants to go himself, Mr. Matthew organizes an overnight trip to the middle and south end of Kruger National Park, thirty miles away. He's anxious to visit before the rainy season sets in and most of the park closes for the duration. He invites me along together with the wife of the cattleman. Mrs. Van Vechmar is overcome with joy at the prospect; she is amazed that the Big Boss should think to ask an employee's wife to go on a wonderful trip like this. It doesn't surprise me, knowing how kind Mr. Matthew is. Later, it occurs to me that he wanted to go and was happy to have me for company, but that such a twosome would look immoral to his neighbours so he invited a third person as well.

We set off on Friday afternoon, arriving at Skukuza Restcamp just before six o'clock, after which time the gates are locked and latecomers fined. For reasonable fees, visitors can either bring their own camping equipment or rent cabins for the night, along with free use of hot baths and showers. For meals, there is a restaurant; patrons can bring their own food and cook it themselves, or they can hire a native "boy" to do this. Mr. Matthew arranges for a native to do our food preparation.

After Mrs. Van Vechmar and I leave our bags in our shared *rondavel* (a round thatched hut with walls of mud and wattle), we go to Mr. Matthew's *rondavel* to eat. We loiter over sundowners of orange juice and gin on the *stoep* under a full moon, with a fresh breeze blowing up from the Sabie river; soon, a native "boy" ceremoniously arrives with a meal of hot spaghetti and wieners (from cans), followed by (canned) peaches. Wonderful!

Before turning in, I go to the front desk to buy post cards. A woman waiting there for service looks at me and starts.

"I recognize you," she says.

"I doubt it," I reply, because I know nobody in South Africa besides the people in Grahamstown and at Fleur de Lys.

"Yes. You were driving alone toward Johannesburg a few weeks ago. You had a Grahamstown licence."

"You're right," I retort, astonished. I guess a woman driving a long distance alone really is unusual.

We are packed, fed, and away the next morning when the gates open at sunrise. Kruger National Park, founded in 1898, is huge, 180 miles long and covering an area of eight thousand square miles criss-crossed by a network of unpaved roads along which visitors cruise; the speed limit is twenty-five mph so that game isn't disturbed, and the dust is kept down. Mr. Matthew drives two hundred miles during the day, Mrs. Van Vechmar and I peering out the windows and calling the names of animals we see, but we cover only a small part of the park. Luckily it's near the end of the tourist season so there are few other cars about.

I'm thrilled at the many giraffe, but since we aren't allowed to leave our car or drive off the road, they are far less accessible than giraffe at Fleur de Lys and so I soon focus on other species, too. Most numerous are the impala— many thousands of them. We pass a herd at least every five minutes, some on the road so we have to slow down or stop. Eventually we ignore them, not bothering to mention a group of a hundred or so feeding nearby.

The next most numerous animals are zebra and wildebeest, lying, walking, and grazing in the thousands; what god of harmony has inspired these species to be so compatible? To jostle side by side in a friendly way at water pools? To lie down contentedly a few feet apart?

"Wildebeest are the same as gnu," Mr. Matthew says out of the blue. "Gnu is the only Bushman word that's been incorporated into the English language." He thinks a minute, then says reflectively, "but of course we call them wildebeest, not gnu."

Less common, but soon not deemed worth pausing for, are waterbuck, kudu, steinbok, baboons, warthogs, and duikers. The exceptions are a rare red duiker and a steinbok with her baby, who is suckling at the road's edge. Since the mother is only two feet high, the baby is that much smaller, with tiny hooves. The mother is terrified to have a car parked so close, but the baby has no intention of giving up its meal. Later we stop for a jackal and a hyena, the latter so close to a car in front of us that the driver could have touched it.

On a more northerly, unfrequented road in the afternoon, we see six big elephants and two babies which thrill us to the core, or at least thrill Mrs. Van Vechmar and me, since Mr. Matthew has seen lots of elephants before. They're standing in the shade of trees about a quarter of a mile away; one takes a dust bath, blowing occasional geysers of earth into the air.

We reach the Orpen Gate which is on the western side of the park and nearest Klaserie, at dusk, just before it closes. We're overwhelmed with the wonder of the day and only a little sorry not to have seen lion when there are about two thousand in the park. As if aware of our lion comments, when it's fully dark a young lioness crosses the road in front of the car about five miles from the gate.

"Stop!" Mrs. Van Vechmar and I scream.

Mr. Matthew brakes and drives backward toward the lion, who crouches beside a tree newly pushed over by a feeding elephant. When we shine our headlights on her, she hunches lower, wondering what on earth we want. She looks as if she will stand for no nonsense, as if she has no sense of humour at all. We see another pair of lion eyes in the bush behind her. When Mr. Matthew drives forward a little, she follows behind us.

"Look at her muscles!" Mrs. Van Vechmar says.

"She's in superb condition," agrees Mr. Matthew.

"I hope no African comes walking by," I say.

"I'd worry about her, too," says Mr. Matthew. "Lions are classed as vermin outside the park. Farmers shoot them on sight to protect their livestock."

"Have there been lions at Fleur de Lys recently?" I ask. Oh no, I think, now that I've seen a lion in the flesh; here is something else besides snakes I have to worry about.

"Sometimes," he answers. "We had one last year that killed a young giraffe. I'll shoot them on Fleur de Lys if I have to."

"It's scary," I comment, "to be a prime attraction in one place and shot on sight a mile away. What can a lion make of that?"

"What indeed," says Mr. Matthew.

Although I'm present every day on the farm and spend much time while eating and talking after dinner with the big boss, Mr. Matthew, for some reason I'm kept out of the loop about what is actually going on day to day at Fleur de Lys. This can be a problem. When I visit one paddock on a late afternoon, I'm horrified to find the whole field on fire. Not only will the grass be destroyed, but the giraffe I can see through the smoke, one a baby, may be burned. I speed back in Camelo to the farmhouse to warn Mr. Matthew, feeling like the boy who saved a Dutch town from flooding by putting his finger in the dike (rather an unlikely story, given the size of a boy's finger).

"Fire, fire in the second paddock," I cry, running into the office where Farnie and Mr. Matthew are going over sales figures at the main desk.

Mr. Matthew looks startled, then embarrassed. "Oh," he says. "I should have warned you. I had the boys set a fire there to kill some of the thorn trees. The rainy season will soon be here and then we'll have fresh new grass for the cattle."

"Oh. Sorry for the panic," I say, "but there are baby giraffe in the field. Will they be all right?"

"They'll be fine," he says. "It's just a grass fire. Giraffe can step over the flames."

Mr. Matthew stops work and drives over with me to the paddock, ostensibly to check up on the fire but also perhaps to make me feel better. The cattleman is there now, trying to shoo two female giraffe and a baby away from the fire, which now surrounds them on three sides and seems to me to be out of control, with flames leaping ten feet into the air. A burning tree is transformed into a flame-thrower, the wind blowing red embers in a large radius from its top onto the grass beneath. The cattleman, who is behind the giraffe and also in the arc of fire, waves his arms at them to try to move them along the passageway the fire hasn't yet co-opted.

"Get out, get out," he shouts, but they stand still, frozen in fear.

"That way, you stupid brutes," he yells, adding a few choice swear words.

Only when the fire almost touches them do they break away in a rush, sprinting through the flames to the burnt grass beyond. Once they are away from the fire they stop running, apparently forgetting their close escape. The baby, so new (although six feet tall) that it has part of its umbilical cord still attached, begins to nurse from its mother. She puts her head down gently to touch its back. Soon the mother walks toward the fence marking off the adjoining paddock and steps over it. The baby, close behind, is much too small to do the same. The mother strolls off into the new paddock but then returns to the fence, staring at her baby, wondering why it isn't following her. They stand like this for a few minutes looking at each other; then the mother turns and strides into the bush, leaving her young behind.

"Doesn't she realize the baby can't get over the fence?" I ask Mr. Matthew in anguish.

"Giraffe aren't very smart," he admits.

"What'll we do? Can we cut the fence?"

"No," says Mr. Matthew. "That would scare both of them away and, anyway, it's getting dark. We'll have to leave them. Probably the mother will come back later."

That evening, while Mr. Matthew is regaling me with stories about his daughters, I can think of little but the baby giraffe, maybe alone, in the burnt paddock. What if a lion finds it? A leopard? A hyena? I feel sick with worry. The next morning I rush back to the paddock at dawn but can find neither the mother nor the baby. We never discover what happened to them—another crisis dissipated.

As I'm driving home for lunch that day, I pass Mokkies on his bicycle and stop. When he pedals up to Camelo, I stick my head out of the car and grin. He smiles back.

"Any giraffe records yet?" I say. He doesn't have the wooden plaque around his neck, but then he may be on some other errand. I point to him, stretch my arms up as if outlining the shape of a giraffe, and pantomime writing on a plaque.

He smiles again and nods his head. What does this means? I pause for a moment, then wave encouragingly and drive on, telling myself that he hasn't had much time yet to carry out giraffe reports.

# 7

## *October*

By the time October rolls around, I feel at home on Fleur de Lys. There's an air of relaxation on the farm now that a few hard rains at night have decreased the danger of fire. I know most of the natives who work around the house well enough to exchange grins, and try actually speaking with Mokkies whenever I see him, although with ill success; I still don't know if he has collected any information on giraffe or on the food they eat. I chat to the white managers I encounter and would be happy to meet their invisible wives, slightly surprised that none has asked me to tea. I never do socialize with them although they all live in nearby bungalows. With the exception of Mrs. Van Vechmar (whom I never see again after our wonderful trip to Kruger National Park), do they disapprove of my living at Mr. Matthew's house? Have they heard rumours that I actually seem to like natives? Do they want to steer clear of Mr. Matthew's attention? Are they envious of my freedom as a woman to do what I want? To come and go as the spirit moves me without the constraints of children and housework (although they have "boys" to do the actual work)? I'll never know.

I continue with my daily routine to find out what giraffe eat and what they do during the course of each day, a topic on which I am now swamped with data, given that watching ten giraffe even for an hour produces 120 records. I also know, more or less, the social tendencies of various individuals. I came to Africa thinking that the giraffe was a herd animal, like zebra, since so many photographs show a number of them close together, but I'm beginning to wonder about this when individuals are often alone, or with only one other companion.

At first, when I saw a group of giraffe, I tried to determine which individual was the leader, as if there *had* to be a leader. Was it a male or a female? If a giraffe walked away from the herd, would the others follow and so confirm his or her status? Sometimes Star set off on his own toward a new feeding ground or to visit a waterhole, but rarely did others trail after him. Often a female did the same thing, but in this case occasionally a few other females and young went too. My notes soon showed that no male led any group and that although females and young were more likely to stick together, no one individual seemed to lead them.

Then I begin to wonder—do giraffe even live in herds? Maybe they are essentially solitary with a few seeming to hang out together only because they all prefer to be near a certain waterhole or a good feeding area. Often the animals I see are spread over a large area and pay no attention to each other, walking away from colleagues if they feel like it, or moving from one area to another to join a second group browsing at a tree. Star is with six giraffe on one day, with two others the next, but all alone on other occasions; Pom-Pom is sometimes with Pim, but sometimes also with the twins (two females I can't tell apart), or with Lumpy or alone.

Looking back from the present, it is embarrassing to have to admit that although male and female giraffe are not usually together in the wild, my mind fifty years ago was determined to envision "the giraffe family." This fixed concept began when I was twelve, suffering from scarlet fever and incarcerated for a month in the Isolation Hospital in Toronto. My mother kindly made me a stuffed giraffe with a pink ribbon around her neck to keep up my spirits. Because I loved this animal so much, and was desolate to learn that since she was germ-ridden she would have to be destroyed rather than return home with me, my mother made me two more giraffe, one with a blue ribbon and the other a baby; these two would welcome me home. However, the hospital staff, who had also noted my love of the stuffed female giraffe, made a special effort to have her sterilized so she could live on. So, when my school life resumed, I gloried in a male, a female, and a baby giraffe. The concept of the giraffe family was born for me—a mother, a father, and the baby they had produced. Unfortunately, and unscientifically, I would write explicitly about the giraffe family in my scholarly monograph of the giraffe, even though the evidence for such a family was non-existent.

To return to Africa in 1956, when I drove off each dawn I never knew which individual giraffe I'd find together. I would make notes about where the nucleus of the Game Reserve giraffe had been located the evening before, and usually there are several whom I recognize in that area the next day. I am able, therefore, to estimate *very* roughly how far the animals in the Game

Reserve travel each day, which isn't far at all. The collected information for the next three months shows that an amorphous group wanders, on average, less than a mile from one dawn to the next, most of this at night. I've read that giraffe never cross rivers, perhaps because they can't swim (one captive giraffe drowned when it was accidentally dropped from a crane while being swung on board a ship); however, they often cross the shallow Klaserie River to browse on the far side.

Giraffe do have some sort of herding instinct in times of stress, because twice groups of strange giraffe appear on Fleur de Lys for a period of a few days before moving on. In September, thirteen emaciated animals arrived with necks so gaunt that I could see the placement of each of the seven vertebrae, as if the skin covered a chain-linked structure. Mr. Matthew thought they were from the north, where the winter had been especially dry.

About the same time I saw eleven giraffe on the west side of Fleur de Lys, where I'd been told giraffe were unknown because of the many fences and heavy farming. They had blackened coats, as if they'd been forced from their usual haunts by fire. They were more nervous than our own giraffe when I stopped Camelo to watch them; the next day, they too were gone.

After a month at Fleur de Lys, I witness Major Hostility for the first time. It's about seven o'clock and I'm parked on the road beside the Guernsey Area watching Star and at least three other males near the fence. (There may be more giraffe in the area, but they're too well camouflaged to detect among the scattered bushes and trees.) Two of them are sparring, one is browsing on an acacia tree, and Star, about thirty yards ahead of me, is staring across the road at Paddock 7. When I follow his gaze, I see four other males hanging out there, looking at nothing in particular, neither feeding nor sparring, which is unusual. Soon I notice that one of these males (whom I'll call Dark because of his coat colour) is staring intently back at Star. Both ignore Camelo, although she isn't far away. Star stalks right up to the fence and, a moment later, Dark moves forward to his fence too. They are now about thirty feet apart, separated by the two fences and the road and "glaring" at each other.

Suddenly Star seems to win this intense psychological battle, if this is what it is, because Dark pivots and canters back fifty yards from the road. Star turns and stalks toward his male colleagues in the Guernsey Area, who immediately retreat before him in disarray, the two males halting their sparring session and the third galloping away into the bush. Did they observe the interaction between the two males and sense that Star is in aggressive mode? Is it something about his carriage that stimulates their unusual reaction?

The excitement isn't over, though. Star continues to glance over to Paddock 7, where Dark remains well back from the fence. The other three males, however, have approached their fence so that Star returns to his and begins staring at them, rather than at Dark, after first stretching his nose high into the air. The three back off somewhat but then one of them, a light-coloured bull whom I'll call Blond, steps bravely over the fence, crosses the road, and approaches to within touching distance of Star, with only the wire fence separating them. As if following a script, they both turn away from Camelo in tandem, walk side by side in lock step for thirty yards and stop, still looking straight ahead.

The next moves are startling. Star spreads his legs apart to give himself a good stance, then whips his head back and gives Blond a terrific blow on the chest. I can clearly hear the impact from Camelo. Blond staggers back into the road, but is brought to his senses by the approach of a car. (Damn.) He steps over the fence into Star's paddock and the two bulls edge into the bush, shoving and pushing against each other, and occasionally hitting out with their heads. After they've disappeared, I count seven giraffe nearby in the Guernsey Area walking about in single file and then in circles, apparently upset by what has happened. If Blond was challenging Star as a primary male, he seems to have been unsuccessful because Star remains in the Game Reserve area for the rest of my stay, and is the only male I see mating with a female.

During October I watch other serious fights, too, which differ from sparring because they take place only between two adult males, require the opponents to spread their legs wide for stability when giving blows, and result otherwise in a great deal of movement. The giraffe act rather like fencers, with much advancing and retreating between wallops. One pair circles a tree at least ten times, whacking the rump of their opponent as they stride. Sometimes, after blows, a bull stretches his nose into the air as Star had done. Is he trying to appear taller and more formidable? Some serious fights end in disaster. A local farmer, Mr. Grant Wiggell, tells me he saw one giraffe knocked senseless in a battle; he lay unconscious on the ground for twenty minutes before struggling to his feet and wandering off in a daze. Other giraffe have been reported killed in such combat.

Ferocious as such fights seem, they sometimes deteriorate at the end into sparring matches which, to me anyway, undercut their apparent violence. One fighting male loses interest in the match well before his partner. Rather than continue to hit his opponent he tries to "neck" with him instead, rubbing his head along the other's neck. The opponent responds with a volley of blows that forces his partner to desist and retreat. It is many minutes before the peacemaker manages to calm down his adversary.

The day after Star's fight, I drive down the Guernsey Road at first light to see if the male giraffe are still about, but I glimpse only Dark, too far back in Paddock 7 to be worth my attention. I head for the Game Reserve to see if Star is near the circular track, but there's no sign of giraffe there. I do spy the resident herd of sable across the Klaserie River, the one from which poachers have been shooting males. I try to count them to make sure there are twenty-two, that poachers haven't killed any more, but there are too many lying beside and behind each other to make my count accurate.

I sweep the far Guernsey Area with my field glasses to ensure there aren't any giraffe there either, when to my disbelief I see another herd of sable on the grassland. We have somehow doubled our number of this rare antelope! I drive back to the farmhouse in euphoria.

"There are two herds of sable in the Game Reserve!" I cry to Mr. Matthew as I rush in to breakfast.

"There can't be," he retorts. "Mokkies and his boys would have reported them. Probably some cattle got into the field. They have long horns too."

"No, I'm sure they're sable. They have straight horns sweeping back."

After breakfast he drives out with me to see for himself. There the two groups are, their members still grazing or resting about half a mile apart. There are obviously many more than the twenty-two sable expected—later we count thirty-four animals in this new herd. Mr. Matthew is thrilled. We drive immediately to the west side of the farm, where he has donated sixty acres of land for a provincial game warden who has built a house there. The game warden too is elated.

The next week, just before lunch on Friday, this game warden comes to the house with an upsetting message. "I've had to shoot a big male giraffe on the main road," he announces. "He was frightening the donkeys and a danger to traffic. I couldn't get him into one of your paddocks."

I'm aghast. How could anyone shoot a giraffe! There are hardly any donkeys *or* traffic on the main road!

Mr. Matthew is practical as always. "Can we have the meat?" he asks.

"Yes, and I brought you its tail as well, in case anyone stole it." He hands over to Mr. Matthew the bony root of the four-foot-long tail with its thick black hairs, rather like those of a black horse.

"The giraffe's about a mile from here," he continues. Thank goodness. That means it won't be Star or Lumpy or the other males I've seen that day in the Game Reserve. Maybe it's Cream though? I haven't seen him in weeks. I feel sick with worry.

Mr. Matthew gives orders to Farnie that a truck, twelve natives, and two

managers be sent to bring in the carcass. He, George, and I don't talk during
the sad main course of lunch, but over dessert I gather up my courage.

"Can I measure the organs?" I ask Mr. Matthew. This seems like the sort
of thing a zoologist should do, and any information I can collect from the body
will make the giraffe's death seem less awful.

"If you want. You can work behind the milk shed this afternoon."

I pace around the house for the next hour, but the truck doesn't return.
Then a worker rides up on a bicycle to report that the body is too heavy; even
fourteen men can't wrestle in onto the vehicle. Oh no, I think. Cream was
huge.

"They'll have to cut it up on the road," Mr. Matthew decides. "Anne,
I'm busy here this afternoon. Can you take them a set of knives? Use my
car."

I wince at the thought of knives ripping apart Cream or any other giraffe
but agree, remembering that I'm a zoologist. The truck is parked at the side
of the road in front of the carcass, with two planks leaning against the back up
which the men had tried to push the body. The men are now sitting around
on the ground, smoking and chatting. I look at the carcass carefully. The
brown spots rest on a white background of skin, so it's not Cream! I breathe
a huge sigh of relief.

"Here are the knives," I tell Van Vechmar as I hand their case over to
him. "Mr. Matthew says you'll have to cut the giraffe up out here. He says I
can have the organs to study."

"Thanks," he replies. "You really want the insides? There'll be a lot."

I nod. He rolls his eyes, then orders his men to get up and start work again.
First, a ranger in uniform cuts the tough skin along the stomach, down each
leg and up the front of the neck. Then the others yank back the skin at different parts of the body to reveal red muscle and white connective tissue.
They sing tuneless songs as they tug and hack at the flesh with their knives;
when it comes time to turn the body over so they can work on the other side,
they join in a single song to coordinate their efforts. The skin, over half an inch
thick on the torso, has many ticks of three species, which I collect in a small
jam jar. Afterwards, Mr. Matthew allows me to add whisky to the jar as a preservative and I'll have them identified. Each is a known carrier of a disease that
affects cattle, but no one knows the susceptibility of giraffe or other game.

The male is such a large animal that it takes the men three hours to hack
its carcass into manageable pieces. Even then, the flesh will have to be divided
into much smaller portions the next day so that all the workers and their families can have a share. The stomach needs six men to lift it into a basin and onto
the truck; around it they dump the endless lengths of slippery intestines. The

twenty-five-pound heart fills a large bucket. The road is like a butcher's shop, with men staggering about carrying bloody segments of leg, or liver, or spleen.

It's upsetting to see the men hack the giraffe's neck into lengths with an axe, as if it were cordwood; they chop along the neck at equal intervals so that the seven vertebrae are separated, each surrounded with its casing of muscles, nerves, and ligaments. Even more upsetting is the head of the male sitting upright on the road, his eyes dull. I go over to inspect it. Once it's cleaned of flesh, as I learned to do at the museum in Toronto with fox skulls, it will be useful for identification of this subspecies of giraffe, *Giraffa camelopardalis wardi,* later changed to *G.c. giraffa.* I try to lift it but, awkward as it is, I can't even hoist it clear of the ground. I'm amazed at its weight. Finally Mokkies and one of his men lug it to the truck. (I never do send it, or any other specimens, to the Royal Ontario Museum in Toronto. The Curator of Mammals had suggested I do this, but he must have been in jest because he organized no way in which their transport could be paid for, and I don't have money for postage.) Years later, when I visit various museums to study the skulls of both male and female giraffe, I find that the adult male skull, at about thirty lbs. (without skin, muscle, or other organs), is three times as heavy. Whereas the female skull has many sinuses inside which keep it light, the male's is made of more solid bone, which enables him to pack a huge wallop in fights with other males. His horns are also larger, and their impact in fights increased by the addition of other bony growths on his face.

I spend all Saturday studying my haul by the milk shed in a haze of flies, blood, and rotting smells. First I spread out the intestines so I can photograph them and measure them with a yardstick. They're about two hundred and fifty feet long, reflecting the giraffe's vegetarian diet. (The tourist who apparently asked staff in Kruger National Park if the giraffe "didn't take a heavy toll of the poor little antelope" would be relieved to learn this.) I had hoped to find bits of leaves in the stomach contents that I could identify and add to my food list, but this is impossible; I'm confronted instead by gallons of juicy green liquid. I know that giraffe have four separate stomach areas, as do cattle and other ruminants, but, for lack of an anatomy text, find only three. The biggest area, the rumen, holds about a hundred pounds of small twigs and "guck." Another chamber is smaller, with ridged walls. The smallest, which contains the finest green particles, is either the omasum or abomasum, the area to which food is returned after being chewed as cud and before being passed along into the intestines.

The heart is as long as from my fingertips to my elbow, and the aorta much bigger in diameter than a silver dollar. The blood is sloshing around by the gallon. The natives usually consume the blood, but there is so much meat

Cutting up a giraffe for its meat.

The lengthy intestines of the large male giraffe.

this time that they discard it for want of space and refrigeration to store it. As I'm turning the heart over in the bucket, I see a small, six-inch snake in the dirt nearby. I point it out to Jessie, the dog, who has been watching my activities with interest, but at its next wriggle she's off like a shot towards the house.

As I'm completing my studies in the afternoon, men leaving work ride by me on their bicycles, clutching five-pound chunks of giraffe meat under their arm. I glimpse Mokkies with a section of neck wrapped in newspaper and call out to him, but he doesn't hear me. I still have no idea yet how his research on giraffe is going.

A close friend of the Matthews, a widow named Mrs. Cook, arrives to stay with us for a while. At the time I think nothing of it, imagining that Mrs. Cook is simply visiting, even though she and Mr. Matthew have little in common. But was this a family emergency? Did the Matthews invite her because of worry about gossip based on my presence? I had already been confronted by a farmer I had never seen before when I visited the tiny store in Klaserie.

"How's your husband?" he had said.

I looked around to see if he was addressing someone else, but we were alone, except for the Indian shopkeeper.

"I'm not married," I said, confused.

He gave a bark of a laugh and turned away. At the time I thought he had mistaken me for someone else, although this seemed odd when there were so few white people in the area. Later, I realized that he must have been referring to Mr. Matthew as my husband. Were the other farmers shocked at my living in Mr. Matthew's house? Were they envious of Mr. Matthew's highly prosperous farm? No matter—I would have ignored them at the time no matter what they thought, especially when their concern about Mr. Matthew's morals were unfounded.

Mrs. Cook is not only friendly but loves to bake, even though she must do so on a wood stove. The dessert situation is much improved. Whereas Watch only makes three desserts which he rotates every third day—jelly, rice pudding, and instant pudding, rather like my own repertoire back home in Canada—Mrs. Cook delights us with an interlude of pies. Watch is no longer head chef but he doesn't seem to mind, and neither do we.

"Watch is a miserable little creature, but it's amazing what he can do with his twisted arm," she tells me after dinner one day. "He can be most annoying at times. You have to treat the natives fairly, but always remember that they are just like children."

I start to object but she cuts me off. "You don't live with natives, so you don't know anything about them. They don't appreciate anything you do for them, and Mr. Matthew has done a lot. We must always hold ourselves superior, both in dress and manner, always." She gives me a look. I always wear a cotton skirt and blouse for dinner (all regularly washed and ironed by Nelson), but I imagine she's thinking about me dressed in my unfeminine khaki field uniform.

"If you knew some educated Africans and met them socially, they'd seem different," I say, thinking of Josiah.

She gives a humph of annoyance. "Education just gives them a superficial whitewash. I have a friend who knows Father Huddleston well, even though he entertains natives. My friend wants to bring some of Father Huddleston's friends to my house so I can meet them. Of course I would never have that."

She pauses for a second piece of pie. "Indians are no better," she continues. "They're filthy as well, but smarter. They've cornered all the rice in South Africa. Just like them."

I decide that I can't talk about workers with either Mr. Matthew *or* Mrs. Cook.

Some days later, I find she doesn't like Jews either.

"Sometimes I wonder, with all this fighting in the near East, if the Jews really are the chosen people," she says dreamily at dinner. "The Jews do give a lot of money to Christian charities, but of course they wouldn't do this if it didn't benefit them too."

"It isn't fair to group people together like that," I say. "Everyone's different."

"I know what I know," she retorts smugly. "When my husband George was alive, people used to call us George and the Dragon."

After dinner, Mr. Matthew gives Watch and Nelson not only their weekly bonus of a package of cigarettes and matches, but also a dog biscuit each as a joke.

Watch laughs and says "Jaas, boss," putting the biscuit in his pocket. He *really* laughs, along with Mr. Matthew, when Nelson starts to eat his biscuit. When Watch tells Nelson that it's a dog biscuit, Nelson chokes, then laughs too.

Kruger National Park will soon close most of its area because of the approaching rainy season so Mr. Matthew and I, with Mrs. Cook this time (another covert chaperon no doubt), make a second trip to Kruger National Park to see the north end, the domain of elephants, which we had no time to visit with Mrs. Van Vechmar. On the way to Letaba Restcamp our appetites are whetted by the sight of thirteen elephants, three of them only an elephant's length away. We eat dinner overlooking the Letaba River, watching an elephant munching reeds on the sandy bank below. Mrs. Cook, who is quite stout, has brought along delicious, home-made chocolate chip cookies for dessert. As I stare through binoculars at the elephant I hear her conversation with Mr. Matthew, larded with "I really shouldn't...," and "They're just too good to resist...," and "Well, just to keep you company."

The next day, as we set off at five o'clock, we encounter two huge elephants just outside the camp, strolling toward us over the crest of a hill, flapping their ears. They step off the road when they see us, but only by fifteen feet; I can see only their legs and stomachs from the inside of the car as they pass. (I now have the back seat to myself, with Mrs. Cook beside the driver.) They are so used to vehicles that they soon stop walking and begin feeding. Mr. Matthew turns the car sideways for a good view of them, then turns off the ignition.

"Shouldn't we leave the motor running in case they attack?" Mrs. Cook asks faintly. This is what most visitors do.

"No," says Mr. Matthew, fiddling with the camera lens. "When the engine's running, the car shakes and ruins the film."

He takes movies of the pair, I photograph them with my camera and Mr. Matthew's other camera, and Mrs. Cook sits in a state of nerves because she's terrified of elephants and there is nothing she can do to protect herself.

"Don't you think they are a little close?" she asks several times in a small voice.

"They're fine," Mr. Matthew insists. She would rather he insist that *we* are fine.

These are only the first of about a hundred elephants we encounter on the way north to the Limpopo River, the boundary of South Africa with Southern Rhodesia, later to become Zimbabwe. I'm agog to see "the great grey-green, greasy Limpopo River, all set about with fever-trees" made famous by Kipling. To my surprise, far from being great, green, and greasy, it's a mere trickle of water in a sandy riverbed because of the summer drought. The fever trees are there though, a species of acacia with yellowish leaves so named because it grows in wet areas where malaria is common.

Luckily for us, the Pafuri River, which runs into it at this point, has pools of water alive with hippos. There we meet Kirk, an African ranger with a gun. If he accompanies us we're allowed to leave our car, so the four of us troop down to the water. The shady river bank is magical, with hippo eyes and ears before us and antelope all around amid enormous trees with thick lianas and creepers hanging down. It's quiet except for the snorting of hippos and the bark of antelope.

As we watch the hippos rising and submerging, suddenly there's a commotion on the opposite bank. A bushbuck, chased by a native dog, comes bounding out of the bush. With nowhere else to go, it jumps into the river and swims toward us; the dog stays on the bank, retreating into the bush before Kirk can shoot it.

"Dog belong to Portuguese poachers," Kirk explains in halting English as he lowers his gun.

The terrified buck has reached our side of the river, but can't climb the bank near us because it's too steep and covered with brush.

"Can we help it?" I ask Kirk anxiously. He shrugs.

We spot a crocodile gliding below our bank toward the splashing animal. I see the bushbuck's huge eyes looking up at us. Then, with a splash, buck and crocodile disappear. Minutes later the crocodile surfaces in the centre of the river, swimming away from us with the buck's torso bulging from the side of its mouth. It reaches the far bank and disappears under a clump of dead trees hanging over the water. We're speechless from this drama, which is over so swiftly, and quietly, and grimly. The hippos continue splashing about as if nothing has happened and there is nothing left to show that anything *has* happened. Subdued, we walk back in single file to the car.

Giraffe don't live in this northern area where the vegetation is quite different than that further south, with obese baobab trees, gigantic fig trees, and sausage trees, their name reflecting the hanging, sausage-like fruits. One large tree on the road is singled out by a sign which proclaims it as marking the boundary between Rhodesia, Mozambique, and the Union of South Africa. Because of the danger of malaria in the relatively wet area (although still dry to my Canadian eyes, judging by the almost waterless Limpopo River), there is no restcamp here for tourists and relatively few visitors of any kind. One herd of eight elephants feeding beside the road is so unaccustomed to cars that its members stampede at our slow approach. We do see many ostrich, nyala, and kudu, as well as the ubiquitous impala.

I've found it hard to become friendly with natives, what with all the white people on the farm watching any amicable overture I make to workers with disapproval. Because of this, I'm pleased to find the family of one of the cattle "boys" living beside the Giraffe borehole, where I spent my first day in the field at Fleur de Lys. On that occasion I'd been so enthralled by the giraffe coming to drink that I hadn't noticed any human beings nearby at all. The next time I drive by the borehole, however, a woman with a baby comes out of a small hut to open the gate for me. I climb out of Camelo to greet her and admire her baby, so we become friendly even though we have no common language. With much sign language I think that she asks me where is my husband? And where are my children, since I am certainly old enough to have them. I gesture that I don't have either, to her surprise. I shrug to show her that I don't mind being so deprived, which puzzles her even more. I'm not interested in babies, but I hold it in its blanket for a few minutes and chuck it under the chin.

Sometimes when I feel in the mood and have Mr. Matthew's movie camera with me, I park Camelo near the giraffe trough with the sun behind me, and wait for a giraffe to come and be photographed while drinking. I'm soon joined by two children, Enoch, about six, and Bella, about thirteen, who sit on the running board to talk to me. Neither speaks a word of English, so I bring along a dictionary of Fanagalo I've purchased to translate for us. This is sold as a practical handbook for English people working with natives, with useful sentences such as "I think you are telling lies" and "Clean the bath first and wash your apron later." However, the children often speak their own native language, Basuto, which is something else again.

One day Bella shows me her three "books," all tiny pamphlets of religious quotations in Basuto, and a religious picture neatly framed in glass with an English legend. This picture features old-fashioned sinners, devils with forked tails, and good women in the fashion of the 1890s with dresses down to their ankles. The chief sinner is killing another man prostrate on the ground. What can she make of this? I read English and even I can't figure out what is going on.

When Bella and I talk, she points to her belt and says the word for "belt" in either Basuto or Fanagalo, I don't know which. Then I give the English word and we both laugh because this seems funny. Then we move on to "hair" or "button." Sometimes I repeat what she says, but so badly that she doesn't recognize that the word is supposed to be in her language. She thinks my new word must be English, and copies it for that reason. Then more laughter.

Sometimes Enoch drives with me to try to find giraffe; he's helpful because he can open and shut the gates we pass through without my having to do so. We each speak our own language so communication is limited, but we share a feeling of camaraderie. Once I'm sure he says "four," but when I look that word up in Fanagalo it isn't there. But soon we come to four giraffe.

October ends with another snake scare, despite Farnie's assertion that I will probably never see a second snake after the cobra. I've read so many stories about the speed of the mamba, the lethal venom of the puff adder, and the expectorating ability of the spitting cobra (which I now know all too well), that I scan the ground carefully wherever I walk. Theoretically I stalk giraffe with infinite care; in reality I scuff along at least noisily enough so that any sensible reptile ahead of me will retreat. With this precaution I feel relatively safe from cobras and puff adders, but not from mambas. Mambas are said to be unpredictable and far more lethal than the others, the bite of one killing a person within minutes.

"Never get between a mother mamba and her hole," Mr. Matthew warns me solemnly—advice I would gladly heed if I knew where a mamba was, let alone her hole. Every night I go down on my hands and knees to peer under my bed for snakes, and every morning I pray to some omnipotent being, "Please, please, don't let a snake bite me."

The cattleman, Van Vechmar, does not soothe my fears. Instead, he tells me about the black mamba he and Jackson, the head native cattleman, met while walking through one of the giraffeless paddocks that I don't visit—a mamba that may not even have been a mother. It suddenly reared up in the grass in front of them, its head swaying four feet above the ground. Jackson, leaping backward in horror, knocked Van Vechmar off his feet; together they fell in a heap to the ground where they lay absolutely still while the mamba surveyed them from above. Finally, deciding whatever suspicions it had were unfounded, the snake lowered its head and backed a few feet away. The two men started to crawl backward too, but at their first movement the mamba was on the alert again, towering over them. For what seemed an eternity they remained motionless while the mamba again slid slowly away, stopping every few yards to raise its head and look back at them.

"Although Jackson's a native and black," Van Vechmar ends his story, "his skin was white that day."

"Is that mamba still about?" I ask nervously.

"Probably," says Van Vechmar. "I still don't go into its paddock unless I have to, and you should never go there."

"I won't," I promise, a promise I know I'll keep.

This account strikes me so forcibly that when I chance upon a fresh mamba moult in the veld, I only regain my shaken composure when I've run back almost a mile to Camelo. (I have since read of a man who was killed by a mamba that pushed its head up through the floor boards of his car and bit him on the heel, so even Camelo was not the haven I then considered her.)

I don't have to go to the field to meet a mamba though, because a mamba comes to the house. A native sweeper encountered it late one afternoon in the grass beside the cattle shed, where it reared up to look him in the eye. He turned and ran without it following him, but while other workers gathered up sticks to attack it, it slithered out of the grass and up into the tree shading my bedroom. If the screening had been off the window, it might have slipped under my bed!

When I arrive back at the house after a giraffe watch, about thirty men, women, and children are milling about the tree, not too closely, pointing up at the snake and chattering excitedly. I join the confusion, peering upward through the branches. I can only make out the snake's black head and bright eye looking down at us, an object of pity rather than terror now that it's at bay. I grab Jessie by the collar and drag her away from the tree toward the house.

"It's a snake," I tell her, pointing up into the tree. "Mustn't be too close." She sits down beside me, licking my leg.

When the cattleman arrives with a shotgun, he aims carefully at the snake's head twenty feet above him. With a sudden blast it's all over; the head jumps, shudders, and slumps from the branch on which it had rested; slowly the whole creature slides through the branches to land with a thump on the dusty ground. We all leap back as it falls, then, seeing that it's quite still, slowly edge forward until we surround it. I find myself standing next to Enoch; we make a worried face at each other, then grin.

The mamba is nine feet long when the cattleman stretches it out to measure it, but so thin it seems no more menacing than a piece of rope. Several men kick the snake's body with their bare feet, but they stay away from its head; venom even from a dead snake can enter a cut or scrape and poison someone. Slowly the Africans lose interest in the mamba and drift away to their chores. Whites and natives had all felt so close during the excitement—pointing, discussing snakes, discussing mambas and especially *this* mamba, which came right up to the house, right over where I sleep—that I'm sad to have our feeling of togetherness evaporate. The sweeper, buoyed by the snake-track bonus, takes extra care in the next few days in smoothing the sand around the house and office, hoping to reveal new snake tracks and earn extra money.

# 8

## *November*

A visiting scientist! In my wildest dreams I never thought of myself as that, especially with everyone calling me "Anne" and I calling them by their formal names. Yet it seems like this is what I am when, in the first week of November, after nine weeks of watching giraffe at Fleur de Lys, I drive with Mrs. Cook to Pretoria, where I will spend a week doing research and boarding at the Pretoria YWCA. Not that my importance is immediately recognized, though. My bed is at the side of a wide corridor, separated from the public by a screen. In the dining room, I sit at a small table facing the wall. For the first twenty-eight hours the only people who speak to me are the secretary who showed me my "room," and a friendly passerby who glanced around the screen to say "Hallo" as I lay in bed. In the next room, about fifteen feet from my ear, a woman now and then plays loud hymns.

My main aim is to have the leaves that giraffe and cattle do and don't eat identified from the fresh samples I've collected from common Fleur de Lys trees and bushes, and especially from those marked with my yellow and red ribbons. Mr. Matthew has a cousin who works at the National Herbarium, Miss Verdoorn, who has kindly agreed to do this. Mr. Matthew, who will be driving up to Pretoria on business later in the week, will take me back to Fleur de Lys when he returns.

After the first day, my sojourn in Pretoria is an introduction to the delightful role of "visiting scientist." The people I meet know nothing about me, but are kindness itself. When I introduce myself to Miss Verdoorn, she arranges for me to have tea with her niece, Val, about my age, and Val's grandfather, because she knows I'm alone in the city.

"I've never met a Canadian before," the grandfather remarks when we gather at his house, looking me up and down. "You certainly have a strange accent."

"Tell us about your country," Val says. "I've been to Victoria Falls and seen more fresh water than I imagined there was in the whole world. Is it true there's even more in Canada?" Having lived all her life in South Africa, she's all too familiar with drought.

"Perhaps a million times more," I say modestly. "We have millions of lakes, and millions of rivers, and Niagara Falls, one of the biggest falls in the world. I live near it. In the winter the lakes turn to ice so you can skate on them, and you can ski down hills covered with snow. We have lots of polar bears but they're all in the far north."

"Do you mind the heat here?" she asks.

"It gets really hot in Canada too. Sometimes as much as 100 degrees Fahrenheit in July and August. We go to the beach and sweat." We're off to a good conversational start about the wonders of Canada.

I drop into the Transvaal Museum to hunt for information on giraffe. The secretary shows me to a work table where, in a few minutes, a native man brings me tea. There is little in the files, but an official involved in Transvaal Conservation shows me a huge map on the wall on which he traces the seasonal movements of giraffe in the lowveld area around Klaserie. While I'm there I meet the director, Dr. FitzSimmons, a world expert on African reptiles. He invites me for drinks that evening at his golf club, and to his home for dinner and a movie with his family.

"Things must be really bad if we need a girl from Canada to come and study our African fauna," he laughs.

At the dinner table I sit beside his nephew, with whom I get along famously.

"I've just read a book," he says proudly on my being introduced as an academic person. "It was by Neville Shute."

He interrupts our talk at regular intervals to throw a ball through the open window for a small dog to chase; the dog rushes out the door, retrieves the ball and, dodging among the feet of the native servants, brings it back to the nephew to throw again. The nephew is deaf in one ear and when we settle down in the bioscope, his mother arranges that I sit on his deaf side so that we can no longer converse. Is this some sort of a message?

A mammalogist at the museum, Waldo Meester, takes me to the Pretoria Zoo, which features a prize raccoon from Canada and an okapi. The latter is the only close relative of the giraffe, first discovered in the Belgian Congo jungle in 1901. This male is large, with a velvety black coat marked by white leg stripes, and has, like giraffe, short horns (but in the male only), a long black tongue, and a sloping back. He walks with his long legs on either side moving forward almost together, just as a giraffe does. As a visiting scientist I'm allowed into the okapi's pen, where I'm nearly able to stroke his nose. (I've written to an expert on this species in the Belgian Congo, James Chapin, hoping that I may be able to visit okapi there. Chapin has been very encouraging, even working the letters of my name, Innis, into a design of a recumbent giraffe, a not inconsiderable feat, but I find that such a trip will be too expensive for me to undertake.)

The next day Waldo borrows the museum's Land Rover and we drive the short distance to Johannesburg. Whereas Pretoria seems a quiet and beautiful city, with flowering, mauve jacaranda trees lining the streets and handsome, red government buildings, Johannesburg is far brasher and busier. By 2000 it will be one of the most dangerous places in the world—one murder trial had to be aborted because all eight witnesses were assassinated. It is the capital of crime in a country where there are twenty thousand murders a year and fifty-two thousand reported rapes.

That is far in the future. In 1956, Waldo and I meet pleasantly with Mr. Davis, who works in the government's Plague Department analyzing small mammals and their fleas. I'm not sure what the giraffe connection is for this trip, but we have lunch with wine and a good time.

After this trip I visit an Indian-owned shop that advertises colourful African blankets (made in Manchester)—"Pay as you wear"—which I will send back to Canada, foolishly packaging them carefully in mothballs although they are made of cotton.

"Apartheid is terrible," the Indian clerk says to me when he realizes from my accent that I'm a foreigner. "I wouldn't want to go back to India, though, because I don't know anyone there. I probably couldn't afford to keep my car there."

"You're doing well now though," I comment.

"I've saved about £5,000 in the ten years I've been here, and now I'd like to use it to visit Canada."

"Canada's a great country," I agree.

On the Sunday that I am in Pretoria, the secretary of the Transvaal Museum, Mrs. Titlestad, whom I met for only a few minutes when she showed me into the library, drops by the YWCA. She invites me to a family gathering at her uncle's farm near Rustenburg, to which she and her husband, a Major, are going.

"I knew you'd be lonely on Sunday in Pretoria with all the stores and entertainment shut down," she says.

As we drive along she worries that I'm not seeing enough scenery. She takes my head in her hands and turns it so I'm looking out the window. "See? Those hills are *koppies*. And that's a banana field."

The members of Mrs. Titlestad's large extended family are bilingual but speak only Afrikaans. Mrs. Titlestad tries at first to translate for me what they are saying, but this proves awkward when she has no cooperation from her relatives; they despise the English for historical reasons, and especially for the defeat of the Afrikaners in the Boer War and the deaths of twenty-six thousand Afrikaner women and children in concentration camps. Anglophone Canadians are obviously no better.

Because the native house staff has Sundays off, dinner is prepared and served by Mrs. Titlestad's cousins, who then retire to the kitchen to eat their meals. At the main table I try to earn my keep by asking the grandmother if she's called "Ouma," more or less translated as "grandmother," to be like Mrs. Smuts, the Grand Old Lady of South Africa and wife of past Prime Minister Jan Smuts about whom I've read.

My reference to an enemy of all Nationalists evokes a collective gasp of horror. I realize immediately that I have made a gross gaffe because Jan Smuts led the Liberal United Party and all of these people are virulently anti-Smuts. Strangely, my blunder is so horrendous that the group decides I'm too out of politics to be worth ignoring.

"Is Canada really cold?" an uncle asks in English.

"Have you ever seen a polar bear?" an aunt wants to know.

"Do you live in a house?" a niece wonders.

After dinner the now-affable grandfather shows me around the farm, which features tobacco, cows, pigs, horses, and bees.

"We've been trying to get rid of the prickly pears with an imported fungus," he tells me, "but now the fungus is destroying the jacaranda trees along our lane. We've got biologists working on the problem."

Before we return to Pretoria, we have a round of coffee and pie at about three o'clock and later a round of ginger beer, these also served by the cousins. On the way home we stop at the colossal Voortrekker Monument on a hill outside Pretoria, built to commemorate the glorious Voortrekkers (who trekked north to escape British rule) and a victory of these Boers against the Zulus on December 16th, 1838. On that date the sunlight at noon falls directly on the Shrine of Honour in Heroes' Hall, which bears the inscription *Ons vir jou, Zuid Afrika!*—We for you, South Africa. At the four corners of the monument are massive statues of four Boer heroes.

As we drive past a bioscope in Pretoria, Mrs. Titlestad tells me about a recent film she wanted to see, *High Society.*

"It opens with a song by a bloody *kaffir,*" she says, this woman who has gone so far out of her way to be kind to me. "There's a huge close-up of his face and his mouth. I could even see *into* his mouth. It was awful. I stalked out of the bioscope immediately but the manager wouldn't give me my money back."

"I had to go back the next day to see the rest of the movie," her husband says with a laugh. Two years later I receive a thrilled note from Major Titlestad, MBE, EI, reporting that his wife, Jelbie, has been delivered of a lovely baby girl and that both mother and child are well. This baby will hate Africans too, if she follows her mother's example. What a pity for the future of South Africa!

On my last day in Pretoria, I return to the National Herbarium to pick up from Miss Verdoorn the list of plants she has kindly identified. It indicates that giraffe browse on at least thirty-two different species of vegetation, far more, I will find, than they eat in December when the trees are in full leaf, although then too their taste is fairly eclectic.

To my delight, I'm able to meet briefly with Griff, who has come to Pretoria on business. When I tell her what I'm discovering about giraffe, she expresses great interest because, unlike Jakes, she's a mammalogist. She agrees to read my final report on giraffe so I can have her feedback. She is turning into a mentor for me, which is marvellous.

"Shall I just put down everything I find out about giraffe?" I ask, referring to the report that I will write.

"Use Fraser Darling's book *A Herd of Red Deer* as a model," she suggests. "That's based on the behaviour of deer he studied in Scotland. There aren't any scientific books about the behaviour of an African species."

"So I've found out," I comment, thinking back to the deficiencies of the zoological libraries in Toronto, at Rhodes University, and at the Transvaal Museum. Later, I realize that I seem to be the first person to come to Africa to undertake a long-term scientific study of an animal. At the time this doesn't even occur to me, since I am only interested in *my* animal, the giraffe, and have no way of knowing that other people might be interested in studying other species. I never imagine that eventually thousands of zoologists will stream into the continent, intent on earning graduate degrees and scientific papers that will make their careers. Being first does not necessarily pay off, because I was never able to find a permanent job in zoology, unlike most of my successors.

"How long can you stay at Fleur de Lys?" Griff asks. "You said earlier that Mr. Matthew suggested three or four months."

"Yes. I'm not sure what to do then. Probably go up to East Africa and search out giraffe there. One article said that Arusha in Tanganyika is a good place for giraffe. What do you think?"

"It could be," agrees Griff. "Can you stay four months instead of three in the Transvaal? You're getting lots of information on the giraffe there."

"I think so," I say. "Mr. Matthew hasn't mentioned my having to leave yet. In fact, he seems to like having me there. He doesn't have any real friends on the farm, and not many away from it." I want to say that I think my being a girl is ideal in my situation, but don't know if Griff will consider me sentimental, which I would hate.

Because of this conversation I write two letters that afternoon. One is to John Cairns, the brother of an old boyfriend studying in Oxford, Alan Cairns, who has told me that John might be able to help me find a base from which to study giraffe in East Africa. John is a District Officer in the colonial administration of Tanganyika, an unlikely occupation for a man of working-class background from small town Galt, Ontario. He lives in Dar es Salaam with his wife, Beverley, and their baby daughter.

The other letter is to the St. Faith Mission Farm in Southern Rhodesia, whose pastor I had met on the *Arundel Castle*. It seems that giraffe live near this farm, so I suggest that I work there in return for room and board and a chance to watch the behaviour of the animals.

I also send off my weekly letter to Ian to tell him about Pretoria. I am still urging him to come to South Africa to teach or to visit (and of course to see me), but he explains that he would be too far from his mother in Winnipeg and would have to give up his job in Ottawa. He continues to sign his letters "Love" though, which is appreciated.

On my return to Fleur de Lys with Mr. Matthew, we stop at a farm outside Pretoria to buy a cow with a good pedigree to be delivered to Fleur de Lys later on. The farm wife chats to me as she shows me her garden and the swimming pool.

"I work for the United Party," she says. "We believe in social segregation from the natives, which is what we have now, but not in economic segregation, which just doesn't work. The Nationalist Party is so anti-English that it's determined to make South Africa into a republic. Children of Afrikaners have to go to Afrikaans schools, where they're brainwashed into despising the English as well as the natives. It's like the Hitler Youth Movement."

"How awful," I sympathize.

"Now the Nationalists want to lower the voting age to eighteen to increase their political clout. No country but Russia has such young people voting! And what a slap in the face to Africans, who don't have the vote at all!"

This seems a little ironic, given that the United Party, when in power, also refused to give votes to blacks.

"I saw the Voortrekker Monument," I say. "It certainly exalts the Afrikaners."

"It cost millions to build, and stands equally for racial hatred and the insignificance of the English," she agrees tartly.

While in Pretoria I've had a chance to think about the rest of my stay at Fleur de Lys. My main focus will, of course, continue to be collecting data on the giraffe, which I'll carry on as I've started, but I decide I'll also try to find out more about the cattle which share the paddocks with the giraffe and about the natives themselves. I think I know Bella and Enoch fairly well but not any of the workers, even Mokkies, who still has not reported to me any of his observations on giraffe. (I've mentioned this to Mr. Matthew, but he has silenced me by saying off-handedly that all of us are busy and that Mokkies has been busy too.)

I don't have long to wait to learn more about the native workers. During our first week back, at lunch one day, Mr. Matthew is more serious than usual.

"One of the boys has been selling alcohol at the lunch break," he says to George and me as we eat the beans and rice Watch has prepared.

"I think I've seen that," says George. "They come back to work stoned. Is it rubbing alcohol or what? Did they get some European bottles of liquor they're not allowed to have?"

"I don't know," says Mr. Matthew, "but I've found out who the boy is and I've fired him. He's pretty smart, so the rest of the workers are scared of him."

"How did you find out?" George asks.

"One of the boys told me in secret—Farnie wrote down what he said. He was afraid he'd be killed if the seller knew who told on him."

As a response to natives drinking on the job, Mr. Matthew and the rangers raid three of the native *kraals* for liquor on the weekend. I wasn't told about this until it was all over. The raid was a success because they discovered 120 gallons of skokian, a mixture made of rotten grapefruit, green pineapples, sugar and vinegar, with calcium carbide added to form acetylene and give the concoction zip. They also brought back to the office huge forty-two gallon drums and clay calabashes full of fermenting mealie porridge for some other brew.

"They really are a crazy, mixed-up people," Mr. Matthew rages to me after the rangers have left. "They use their food making these poisonous concoctions and then have nothing for their families to eat."

Bella                                                    Enoch

"Is it illegal to make alcohol?" I ask.

"They're allowed to make kaffir beer in their *kraals,* but not to sell it. It has a low alcohol content and is supposed to be nourishing. But not skokian, not this poison."

The next week the natives who had the alcohol are tried by the Native Court at Bushbuck Ridge, twenty-five miles away. (Oh that such a quick justice system were in place in the white world!) The rangers had poured out all the skokian at the raid, so there is no sample to test for alcohol content and, because of this, the accused are released. However, two of the women are fined ten pounds each, or a month in jail, for brewing kaffir beer to sell. One admits to making eighty-nine gallons and the other ten gallons. They pay the fines. The men aren't penalized because it is the women who do the brewing while the men are at work; the men claim they know nothing about the beer which is difficult to believe, given the small size of their living quarters.

The following Saturday night at a beer party, one woman, angry with another, runs up behind her and hits her on the head with a rock, knocking her unconscious. Her friends carry the victim to the office, from where Van Vechmar drives her to the hospital for treatment. Her condition seems so serious that Mr. Matthew asks Mokkies to prepare a statement about the attack in case the woman dies. When Mr. Matthew visits the hospital the next day, however, the woman is walking in the hall, evidently unimpaired except for a few stitches in her head.

"The nurse said that natives have thick skulls," Mr. Matthew tells me matter-of-factly. "Apparently her husband went to get help for her when she was hurt, but when he returned to the party she was gone, along with everyone else. He wandered about for a while asking in vain where she was, then went home to sleep before continuing the hunt this morning."

"Then he found that she was at the hospital?"

"Yes. She was lucky. Before you came, a man was knocked out during a fight and suffered a concussion that still bothers him."

I think I see the woman involved because, two days after her accident, Mr. Matthew asks me to drive Watch to the mission hospital to get a checkup from the doctor. Mr. Matthew has told me that he never lets natives ride in his car because they smell—how could they not when they have little access to running or hot water?—so I ask Watch to drive with me in Camelo. (Ever since Mr. Matthew told me this I've made a perverse point of stopping to give local natives a ride if they are going my way, and urging them to sit in the front passenger seat.)

"No, no," says Mr. Matthew when he sees me getting into Camelo. "Take my car. This is a business expense." Perhaps he's embarrassed by his earlier statement? He motions for Watch to get into the back of the station wagon, which he does. On the way over I chat at Watch, but there is scant communication because we have no language in common. I'm pleased to find that the hospital caters only to natives; when Europeans are sick they have to go much farther, usually to Nelspruit, for medical help. In the corridor I notice a cheerful African woman with a bandage on her head—probably our assaulted victim.

The atmosphere at Fleur de Lys is bleak in the days following the raid. The workers are subdued and George says even less than usual at meal times. Mr. Matthew has to spend a week supervising work in Swaziland but, before he departs, Mrs. Cook arrives for another visit. Has Mr. Matthew summoned her because he's afraid to leave me alone in the house? Or afraid to leave me alone with George? I've no way of knowing. Even Mrs. Cook's superb pies don't cheer us up.

Mrs. Cook has no trouble running the house in Mr. Matthew's absence, given that Watch and Nelson do housework full-time, but she can't prevent the dog, Jess, from being poisoned. Watch finds her dead one morning, lying in the yard. The whole farm is immediately in an uproar. Mrs. Cook can't sleep and spends much of each day crying and moping in the living room, blaming the natives.

"Why would anyone do such a thing?" she sobs. "What will Mr. Matthew say? He'll be angry with me."

"It's not your fault," I say, patting her awkwardly on the arm to try to comfort her. "It must have been one of the workers. Jess didn't like some of them."

"The natives don't care much if a *person* dies, let alone a dog," she wails. "They don't think like us. Back home my cook's mother died and he took five days off to deal with witchcraft and whatnot. Then he came back wearing a black piece of cloth on his arm and expected me to sympathize." She rolls her eyes. I look away and roll my eyes too, but for a different reason.

"Jessie was getting old," I say.

"They've no use for dogs. They even treat their women like chattel."

"I've got to be off to the giraffe," I say to escape.

When Mr. Matthew returns, Mrs. Cook's grief reaches a new peak so that he has to expend more energy in comforting her than in mourning himself.

Farnie, the secretary, sets himself the task of finding the culprit who poisoned Jess. He discovers that Jess has bitten one of the native rangers several times, and that this man threatened her after her last attack. On further digging, Farnie learns that the ranger was fired from his previous job on suspicion of poisoning a man, but there was (and is) no concrete evidence against him.

While driving out to find giraffe at sunrise one morning, shortly after this incident, I see a small black and white animal stagger across the road in front of Camelo, looking very much like a skunk. In Canada, under the same circumstances, I would have driven unobtrusively by the animal although it looked to be in distress. But knowing our skunks don't live in Africa (information from a zoology degree coming in useful), I stop Camelo, climb out, and give chase. After a short skirmish I catch the animal in an old shirt. It's so thin that I can feel its ribs through the cloth.

"You poor little thing," I murmur, nuzzling it gently. "I'll take you back to the house to cheer everyone up. Have you lost your mother?"

I decide it has, and march proudly into the office before breakfast, holding my prize. The reaction to the animal is greater than expected. Farnie and two natives helpers shoot out of the office in a body.

"That's a skunk!" Farnie cries. "Watch out!"

"Does he spray?" I ask innocently.

"Of course," Farnie exclaims. "You have skunks in Canada. You must know that. Take it away!"

I can't imagine that this sweet animal would spray anyone. "They don't like you," I say to him, looking into his worried face. I take him into the yard and find a large, wooden box into which I put him.

Mr. Matthew, who insists on calling the skunk Stinky rather than Thomas, the name I've chosen for him, is no more enamoured of him than is Farnie. Although Thomas never sprays anyone, I remain his only friend. He eats bits of raw meat from my fingers and drives with me in Camelo when I search for giraffe. Once he climbs up inside the leg of my trousers as I'm about to turn off the main road; a native sitting nearby dropping pebbles into a jar to count each car that passes (for which he earns five shillings a day) looks startled as I jerk by him in Camelo, clutching at the animal's tail disappearing near my ankle, and giggling as his fur tickles me under my knee.

"I was hoping that Thomas would cheer us up," I say to Mr. Matthew at dinner one night. "He's a nice pet, although of course he can never replace Jessie," I hurry to add.

Mr. Matthew rolls his eyes. "Stinky is getting big enough to look after himself," he says. "Why don't you let him go while he's in good health? Remember the bat?"

This is a low blow on Mr. Matthew's part. In October, Mokkies had found a brown bat on the ground, too young to fly. For three days I cared for it, giving it milk from an eye dropper as it crawled around my hand, hung onto my thumb with its little claws, and chattered with its teeth. The third night it was so noisy, scrabbling around in its empty shoe box, that I put the box in my bathroom and shut the door. To my horror, the next morning it was swimming weakly in the toilet bowl; that evening, it died.

"The bat was very young," I retort in self-defence.

"What about the millipede?"

Another low blow. I had put a four-inch-long millipede in an open-mouthed jar lined with lettuce so that I could study it, but it disappeared during its first night of captivity, even though the jar was in my locked bedroom. The millipede reappeared several weeks later from under the dresser, its shiny black coat sporting a grey dust bunny. It seemed in good spirits, but I felt badly that it had lived in such restricted quarters. I put it outside on the ground where it departed in deliberate haste on its myriad short legs.

"At least it was still alive," is my feeble defence. "You don't think anyone wants Thomas as a pet?" I ask.

"No," Mr. Matthew replies.

The next morning, when I liberate him on the grass in one of the paddocks, he trundles off at top speed, never once looking back. I follow him to a tree where I leave him busily digging a hole among the roots. When I come back three hours later, Thomas has abandoned the first hole and is part way down a second, curled up, and fast asleep. I never see him again.

So far I've collected a huge array of data about what giraffe do during the daytime, but nothing about their activities at night. Do they continue browsing even if it's too dark to see what they're eating? Do they still come to drink? Do they sleep lying down or standing up? I want to go out at night and sit in Camelo in the Game Reserve to answer these questions, but remember Mr. Matthew's warning about danger. His worry about violence is real because, during my week in Pretoria, an African on duty at the store in Klaserie was beaten up one night by a car full of natives. As the man finished putting petrol into their car, he was hit on the head and tied up. He wasn't badly hurt because he was on the job the next day, but he couldn't (or wouldn't) remember the licence number or what the car looked like.

"Will you go out with me to look at giraffe after dark?" I ask George one evening before Mr. Matthew returns from a Swaziland trip.

"How long for?" George asks in a tone of irritation.

"Just an hour. Please?"

"All right," he agrees half-heartedly, "but just for an hour. I don't want to waste the whole evening." What does he do every evening? Play solitaire? Read a book? Listen to the radio? Surely the giraffe will be more interesting.

First I drive to the water trough at the Giraffe borehole, George clambering in and out of Camelo at each gate to unfasten it so I can drive through and close it again after I've done so. We find a solitary giraffe beside the trough who isn't bothered by Camelo's headlights. I leave them on while we watch her looking thoughtfully toward the Drakensburg Range in the distance.

"Put down that giraffe drink after dark," George insists, thrusting my notebook at me.

"But we haven't seen that yet," I protest.

"What else would it be doing here?" he asks, which does seem a sensible question.

"But we have to *see* it drink," I repeat.

George twists and turns in Camelo's small passenger seat, sighing with annoyance. Fortunately, after about fifteen minutes, the giraffe does lean down and suck up water.

"What did I tell you?" snaps George.

"Now let's see if we can find giraffe feeding," I say lightly, trying to revert him to good humour.

"Of course they'll be feeding," he contends. "Would they stand around just looking at trees all night?"

We come to a small group of giraffe, which is startled by my headlights so I turn them off. We edge slowly toward the animals, barely making out their shapes in the darkness. When I turn off the motor there is silence for a moment, and then the distinct sound of mouths tearing leaves and of chewing.

"I told you so," George announces. "Of course they eat in the dark. Why not?"

"But how do they know what leaves to eat?" I counter.

"Who cares?"

I'm prepared to sit for the rest of the hour in the dark, listening to the giraffe's movements in the still air. George is not.

"We're probably missing giraffe," he says after a while. Then, a few minutes later, "I think we're lost."

"Let's go now," he demands finally. "We've seen all we can see. I want to go home."

I drive back to the farmhouse in silence, pleased to have added more data for my report but annoyed that George is so impatient.

"Thanks for coming," I say rather brusquely as he struggles out of Camelo, noting by my watch that we have been away for only fifty minutes.

"Hmmph," he calls over his shoulder as he strides away toward his house.

Perhaps embarrassed somewhat about his ill humour on this occasion, a few evenings later George drives several of us to a biweekly bioscope (for a white audience only) at a local school twelve miles away.

"It'll be a hang of a treat," he says somewhat sarcastically, using a South African idiom.

We arrive at 7:30, the advertised starting time for *The Great Waltz,* but are the only ones present. At eight o'clock, when about twenty farmers have trickled in, a local farm boy sets up the projector. Before the film starts, however, he finds that it is wound on backwards so he puts a pencil through a large empty film reel and rewinds the film onto it by hand. When the film is finally shown, it breaks at least ten times, once even before the cast of characters has finished rolling on the portable screen. At each break the projectionist has to skip a section of the film which makes it difficult to follow the plot, even though the plots of 1930s films aren't very complex. At the end of the film the audience claps heartily, judging it to have been a great success.

"Well," says George philosophically as we drive home, "what with all the breaks we missed most of the film, which saved us some time."

My photographic triumph for November is a film of giraffe drinking at a water hole, a scenario I've tried now and then over a period of weeks to obtain. Once, when some distance from a water hole, I spied a female with her front legs spread wide and her mouth level with her feet, but when I drove closer I found she wasn't drinking water at all. Rather, she had her mouth on the ground yards back from the pool and seemed to be eating dirt, since there was no vegetation there. Like other game animals, she was almost certainly obtaining salt and other minerals from the soil. I waited around on that day hoping she would eventually drink, but she showed no inclination to do so.

A week later I sit sweating in the heat of Camelo from three in the afternoon until five opposite the pool where I saw my very first wild giraffe. No animal ventures there. But just as I'm about to drive away in disgust, five giraffe parade out from the bush, calmly give Camelo the once over, and bend down to drink. I'm able to film two individuals spreading their legs to do so, and another bending them forward before three red cattle amble slowly between Camelo and the giraffe, blocking my view of them. I hiss urgently at the bulls to move on, move on, but not loud enough to startle the giraffe. After a few minutes' pause they do slouch forward again, leaving me with a second chance to film my quarry. I'm ecstatic at my success. Maybe even the cattle will be useful in illustrating just how tall giraffe are.

Usually I think of the myriad free-range beef cattle on the farm as a bit of a pain, even though their presence is vital to Fleur de Lys' prosperity. They are a handsome, reddish breed called "Afrikander," which is able to survive in lowveld conditions where grazing is sparse much of the year and temperatures high. Their long, curved horns give them a fierce mien that they perhaps deserve, because Mr. Matthew told me that one herd banded together to chase a lion out of their paddock.

My impression of the cattle is quite different, though. Often they lie down in the middle of a track so that I'm unable to drive past them when touring a paddock. They watch me in a detached manner as I drive up but refuse to budge until my bumper is within an inch of them. Then they leap to their feet in a panic and rush away as if they've been attacked. If I leave Camelo to stalk giraffe on foot, they gather around her as around an alien creature, licking her steering wheel. I drive home with my hands sticky from their saliva.

# 9

## December

In the first days of December, my giraffe work is interrupted by a marketing crisis—despite the lack of rain, there's a huge crop of ripe lychees at Fleur de Lys. Lychees, red fruits with hard horny coats, grow on trees like oranges and are sold in the big cities as a seasonal crop. Fifteen men are picking branches laden with fruit for twelve hours each day, day after day. The lychees on their stems go to thirty men and women sorters sitting on boxes near the trees, who trim off the leaves and discard the bad fruit. The good fruits on their stems are then sent in boxes to the huge packing shed, where thirty native packers at tables wrap them gently in tissue paper in groups of twenty-five fruits; then the papers are packed in wooden trays, three hundred fruit to a tray, the trays themselves having been nailed together by another work crew. The trays fit into boxes onto which slats are nailed when they are full. Then the boxes are stamped, piled, and finally loaded onto trucks which will take them to the train. Mr. Matthew has booked space for two thousand trays a week, for three weeks, to go in refrigerated train cars for sale in Johannesburg, and another five hundred trays a week destined for Cape Town markets.

The packers are the elite of the workers because they can count to twenty-five, although they do so with some uncertainty and much recounting; they're impressed by my facility with numbers. I'm drafted one week to join this group when only two hundred trays are finished in one day, leaving twenty-three hundred for the next two days. I work alongside the men for ten straight hours, a monotonous job reminiscent of sorting mail in Toronto at Christmas time. The man next to me starts to sing a song which the others take up.

"What are they singing?" I ask the European head of the packhouse, who has been sitting at his ease while the rest of us count and wrap. He gives a shudder.

"They're singing 'We are packing lychees' in Shangaan."

I'm thrilled to be part of this crew, workers singing as they toil, a feature of many movies. "That's wonderful," I comment. "I guess it makes the work go faster."

The overseer shakes his head. "Not wonderful," he says. "You'll see."

And I do see. Several hours later the group is still singing the same words to the same four-note tune. I think my partner has forgotten that he is even singing.

The white managers are proving themselves completely inept. Several times one man orders the pickers to stop picking, but then there are not enough lychees to pack and the pickers are rushed again to the trees. On the last day, when the trucks are waiting for their loads, Mr. Matthew finds that the sorters are casually picking off twigs and tiny leaves from the lichie stems while the overseers chat together oblivious of this unnecessary effort. He rages about, calling the men idiots, swiping natives on the head, and quickly trebling production.

What do I make of Mr. Matthew's aggressive response? Nothing at the time. It must have seemed natural to me that, if someone did something wrong, he might get a swat. After all, I'd been spanked when I was small. In any case the men hit didn't seem to mind and their friends even chuckled. But, of course, my perspective was naïve. A 2003 report of the South African Human Rights Commission finds that little has changed for black farm workers, and describes the conditions I observed as what they really were. Workers are still being assaulted by their white bosses, women are still paid less for doing similar work as men, and most rural Africans continue to live in a climate of extreme poverty, violence, poor health care, and little education. Only now they aren't necessarily suffering in silence: since the end of apartheid in 1994 over fifteen hundred white South African farmers have been murdered.

We packers are slowing down by ten o'clock when Mr. Matthew learns that the train is two hours late, giving us another hour's time to prepare more boxes. This we do with twelve of us working as fast as we can, the natives smirking to each other to see the Big Boss rushing about in action. By the end of the morning, I'm exhausted from the work and the stress of finishing it on time. The white men are philosophical at having had Mr. Matthew take over from them to get things done, and Mr. Matthew has departed to the station to ensure that the transportation of the lychees to market is in place. Watch begins to serve lychees for our meals and I grow to love them.

The workers receive their weekly pay the morning after this lychee crisis (although no one but me thinks that there *was* a crisis). I stand beside Enoch at the mamba-tree outside my bedroom to watch. The men and a few women line up at the office window, where Farnie sits inside with the work records on his desk, handing out money in pay envelopes owed for the week's work. Everyone is in good humour, laughing and joking to each other, although the process takes some time. Many people have to pay back debts from the previous week, or borrow money for the week to come. The maids, as all the women are called, earn seven shillings for a five-and-a-half-day week; one woman who was late several times during the week loses half a day's pay.

At lunch, I ask Mr. Matthew about the workers' low salaries.

"They don't get much, I know," Mr. Matthew tells me, "but they don't work very hard or get much done."

"It seems to be a circle," I say. "They don't get paid much, so why should they work hard? What does Mokkies get?" I'm still thinking about Mokkies and the observations on giraffe he is supposed to collect for me.

"He gets £8 a month, along with his uniform and ninety pounds of mealie meal," he says. "He also shares meat that's shot, like the giraffe, and vegetables in season."

"Is that about average for the lowveld?" I ask.

"Better than average. Mokkies built his house on the farm, grows what crops he needs, and keeps two dozen head of cattle. We give him soap to wash himself and his uniform. His children go to school free, and if they work for us several hours a day, they get money and their food. His wife can work too, washing or hoeing."

"Why do the women get less than the men?" I ask.

"They do different work," Mr. Matthew replies. "The women and children are both free to work at Fleur de Lys if they want, but we have them do things like washing or hoeing or sorting fruit or sweeping the ground."

"Why don't the white women on the farm work?"

"I wanted them to," he states. "It would give them something useful to do. Several years ago I tried to employ several of them, thinking they'd work harder than the natives. They were to stand at a moving belt sorting out bad from good oranges as they went past, but they didn't last long. The work was too much for them."

"Didn't they like the money?" I ask.

"The money was all right, but they're so used to having a gardenboy, a cook, a houseboy, and a wash girl, that they've never done any real work in their lives. Now natives do all the sorting."

If the white women on the farm are aware of what I'm doing, they would be amazed if not incredulous. With less than one month remaining before I have to leave Fleur de Lys, I'm working harder than ever—taking less time for meals and staying in the field each day until dusk. When I came to Fleur de Lys, it was the end of the winter when the dry conditions made it difficult for animals to find enough food. Now it's summer, with all the trees and bushes in full leaf so that the giraffe can eat their daily fill of perhaps a hundred pounds of leaves in only a few hours. They can afford to be selective; leaves of trees and bushes they consumed eagerly in the fall are now ignored in favour of more tasty treats. All the leaves look equally edible or inedible to me. What makes some more appetizing than others?

I decide to collect and dry leaves from plants giraffe like and don't like to see if there are any chemicals present that correlate with taste. Do giraffe like leaves with lots of starch in them? Or dislike those with tannin? Watch lets me dry them at a low temperature in his oven, although he remains perplexed over why anyone would want to heat up leaves in little dishes and then wrap them up, all crisp, in small plastic bags. I send these bags home by mail to my understanding mother, along with a small sample of marble-shaped feces, collected each month, which I think may contain more twigs in early than in late spring. However, when I chemically analyze these strange materials on my return to Canada, I can find no useful results.

When the giraffe aren't browsing, they generally spend their time snoozing and chewing their cud. They do a lot of this in December. It's hard not to yawn when I see several at noon standing in sun or shade with their necks drooping, their eyes almost shut, and their tails motionless. In the winter giraffe rarely lie down, but now at midday I often see several doing so in the Game Reserve open areas—usually younger animals such as Pom-Pom. She has less trouble than the large males in kneeling with her forelegs so that her neck swings low, then collapsing her hind legs to lie down, her legs either to the right or left of her trunk. It's difficult even for her to get up again, though. She rears her neck back and then lunges it forward with enough force that she can straighten her hind legs and put her hind hoofs on the earth. Then she draws her neck sharply back so that her forefeet can also regain the ground.

While lying down in the daytime, Pom-Pom and the others usually keep their heads erect, but sometimes they fall into a short, deeper sleep by laying their necks alongside their body so that their head either rests on their flank,

The head of a baby giraffe.

as in the young animals, or on the ground, chin down, as the relatively longer-
necked adults do. Each day I take the movie camera with me to the field to film
a giraffe in the act of lying down or getting up, but with no success. The
giraffe refuse to change their position when I'm around, or else move so
quickly that I don't have time to grab the camera and focus the lens, or get up
when I'm looking into the sun, which would ruin the film.

When they're lying down, giraffe in the open areas face any direction,
about twenty yards from each other, apparently impervious to possible attack
by lions since few glance about now and then or act as sentinels. Nor do they
seek shade, even if it's very hot. By contrast, camels lie close together in the
even hotter summer desert, with their bodies straight-on rather than broad-
side to the sun's rays in order to reduce the area of skin exposed to the sun.
Both giraffe and camels on hot days raise their body temperature to reduce the
difference in heat between themselves and the air. This helps them conserve
water in the form of sweat needed to cool themselves and enables them, if necessary, to go for relatively long periods without drinking. Their "fur" also
insulates their body to some extent against heat.

Now that so many giraffe have time to relax rather than feed during the
daytime, I capture footage of them chewing their cud. This involves regur-
gitating a bolus of masticated leaves swallowed earlier and chewing it for forty
or fifty seconds, at a rate of one chew a second (I time many cudding giraffe
to obtain this average) before swallowing it again; this rechewed material is
directed to the third and then fourth stomachs from where it enters the small

intestine. I'm able to film Star's head and neck from the front with a bolus charging up, and later down, his six-foot neck, like a frenetic mole pushing up the ground during its passing.

Later, after I've put the camera aside, I see a tickbird the size of a starling (and actually related to the starling) foraging for ticks on the front of the giraffe's neck. It gives a startled jump as a bolus bulge slides downward along the esophagus under its feet. Another photographic masterpiece missed!

Now that there has been some rain, ticks are more common than before, often sucking blood under the giraffes' stomachs and around the genitalia, where there is little hair. This bothers the giraffe, who rub themselves on any convenient spot—tree, rock, and even another giraffe. Pim, a female, stands over a six-foot tall bush, rocking back and forth to scratch her stomach. Pom-Pom reaches back and nibbles an itch on her flank or foot with her mouth.

Reproductive behaviour is rarely witnessed in giraffe because females take no part in it during their fifteen-month pregnancies. After her young is born, a female comes into estrus every two weeks until she again becomes pregnant. As well, there are relatively few females at Fleur de Lys, so I wonder if I will ever see mating activity.

The most common action is the testing of females' urine by male giraffe, something I'd never heard of before. My inauguration is when Pom-Pom urinates spontaneously and Star, standing near her, leans down his head to collect some of her urine in his mouth. Then he raises his neck and curls back his lips as if savouring the liquid. This behaviour, called flehmen by the Germans who first described it in males of various species, enables a male to determine the hormonal condition of a female. If she's in heat, she'll be willing to mate with him; if she's nearing that condition, it's worth his while to hang around. Later, I see Star nuzzling Pom-Pom's flank and nibbling her tail, which inspires her to urinate and for him to again test her urine. On this occasion, two other males approach to join in the activity without apparent jealousy on Star's part.

I don't know if Pom-Pom ever mated with Star, but I do see him mount another female. It's another hot day, the temperature over ninety degrees Fahrenheit, and as usual I'm wilting in torrid Camelo, making notes every five minutes of what all the giraffe I can see in the Game Reserve are doing and allowing myself a sip of water from my tepid canteen immediately afterward. In the distance I spy Star, another male, and a female walking restlessly about. What are they up to? All the other giraffe are stationary and as drowsy as I feel.

After circling her several times, Star stands directly behind the female, who has pulled aside her tail, raises his head and then mounts her, sliding his front legs quickly forward on either side of her flanks. A few seconds later she gives a short run forward, forcing him to dismount. The second male, engrossed with this enterprise, moves closer to the couple. They're mating, I think in excitement. But where is the romance, the necking and tender rubbing of bodies that occurs in homosexual encounters? Did I miss these?

I watch to see what will happen next. Again Star and the female circle about without touching, and again he stands behind and mounts her without ceremony. Again she gives a little rush forward and begins circling Star. The other male is now closer than ever, within five feet of the couple. Star mounts the female a third time, but on this occasion the other male is hovering so close that Star charges him, chasing him into a bushveld area where I can't see them. The female follows after them, out of my view. I'm blown away by what I've seen, and make careful notes, but I don't tell Mr. Matthew and George about it, nor do I describe it in my letters home. As a zoologist I know how central reproduction is, but as a young woman I'm so embarrassed by the reality that, although I describe it in my report, I can't mention it to friends or family.

The day after this momentous event that I don't mention, I receive a letter from John Cairns suggesting that I come to Tanganyika to stay for a few weeks to see if we can work out how to continue my studies of giraffe in East Africa. With my four-month allotment of time at Fleur de Lys almost up, this seems like a good idea. At the very least I'll be able to see the habitat of giraffe there, and I'll have time to analyze in detail the data I've collected on the Fleur de Lys animals. I've already heard from St. Faith's Mission Farm in Southern Rhodesia that there is no possibility of my working or staying there.

"My best plan seems to be to go to Tanganyika after Christmas and carry on giraffe work there," I tell Mr. Matthew after dinner. "I can book passage on a ship going up the east coast of Africa from Durban to Dar es Salaam. Would it be all right if I left Camelo here until I return to South Africa? Then I can drive her back to Grahamstown to sell her."

He looks startled, as if he has forgotten all about his earlier time limit. I'm sure he's sorry to have me go, although it was he who suggested I stay at Fleur de Lys for at most four months. "Of course you can leave the car. Are you finished here?" he asks. "Will you have somewhere to stay?"

I explain to him about John Cairns's offer.

"Well, we'll miss you," he says. "If it doesn't work out there, why don't you come back here and complete a real film on the giraffe? You have lots of good footage already, but no overall theme. We can make a professional film using my editor and splicer."

"That would be great!" I exclaim. "You wouldn't mind putting me up some more? I haven't really earned my board, I'm afraid. I haven't yet finished the report on what cattle eat that you wanted."

"All the more reason to come back in the fall," he says. "And you have been useful—typing reports for me, driving natives to the hospital, taking me to the airport when I've had to fly to Swaziland, packing lychees. You've helped when we needed it."

"I can never thank you enough for letting me come," I say.

"Not at all," the kindest man in the world responds with a smile. "We've enjoyed having you, and I'd like to have a completed film, the first one ever on giraffe."

Mr. Matthew's dream is to have Fleur de Lys made into a government-sanctioned game reserve so that he can protect all the animals that live on it. Now, men who kill game animals pay only small fines or aren't caught at all. Recently Mokkies and his men noticed a cow standing by herself in the middle of a field. When they investigated, they found she'd been snared by one leg. Rangers kept watch on the cow from behind a clump of bushes until an old man of about seventy called Offis, one of the workshop "boys," approached her, then arrested him. The court fined Offis fifteen pounds or six weeks in jail; he went to jail because he had no money.

"Using wire snares is terrible," Mr. Matthew tells me. "Even if an animal breaks free, the rusty wire still digs into its flesh and kills it later from blood poisoning."

"I guess that's what's happening to Limpy, one of my giraffe," I say. (Nowadays one could tranquilize an individual and remove the wire, but at that time this procedure was unknown.)

"I can see poor people wanting meat for food, but not European poachers!" Mr. Matthew exclaimed.

"They probably killed the sable in the winter?" I ask.

"Yes. They have the trucks to cart off carcasses and trophy heads, and the contacts to sell them. The man who built the Game Warden's house was a poacher. He came back one night after the building was finished and killed eleven buck using a lantern and knife. He paralyzed them by shining a light in their eyes, and then stabbed them without even knowing what species they were."

"That's disgusting," I agree.

In December, Mr. Matthew's wish comes true: much of Fleur de Lys farm is officially declared the P.W. Willis Game Reserve in honour of an early pioneer in the area. For this designation the farm has to be fenced, and signs put up stating that anyone caught trespassing or poaching in the reserve will be fined a minimum of £100. Such a person will be arrested by Fleur de Lys rangers and sent to court. Today, the P.W. Willis Game Reserve continues to exist, serving as a research area for scientists investigating various aspects of wildlife. Mr. Matthew would be thrilled to know this.

Near the end of my time at Fleur de Lys I accompany Mr. Matthew again to Swaziland, this time to the Ross Citrus Estate, a young, two-year-old company for which he is a consultant, near Bremersdorp. Already the company is growing rice, and has a small citrus grove and a citrus nursery. While Mr. Matthew talks to the manager I chat with his wife, a pretty blonde German of my age, twenty-three, who used to be a model. She wears a diamond bracelet and quizzes me about fashion in America, a subject about which I know nothing, to her disappointment. She says her hobby is gardening, but as I count twenty-five natives working on her front garden alone, there is hardly room for her. She plays tennis every day on her tennis court, swims in her large blue pool, and rides one of her horses. If she wants to visit her house in Johannesburg, she can drive there in her Buick or fly in the company plane. She has four indoor servants, but complains to me that it is much harder entertaining here than in Germany.

"At home I used to go to a delicatessen and buy what I wanted, but here I have the bother of ordering whatever we're going to eat and then making sure the cooks prepare it properly."

"They seem like good cooks," I say as I munch a cookie that comes with the coffee a servant girl brings us.

"They should be. I hire coloured girls; they're smarter than natives, so I have to pay them £5 a month," she groans.

Mr. Matthew, the manager, his wife, and I have drinks by the pool before we leave. The wife elaborates on fashion while we listen in silence. She disparages plastic handbags as infinitely inferior to leather, so with my foot I nudge my plastic purse out of sight under my chair.

Mr. Matthew and I spend the night in Bremersdorp at a small hotel which serves as a club for white Swazilanders dressed in knee socks, khaki shorts, and white shirts. They come each evening for drinks, dinner, and more drinks. I insist on paying for my own room even though Mr. Matthew offers; during the whole previous month my only purchase was a bar of soap, so my finances are in much better shape than they might have been.

Beside the small swimming pool an English boy sits with a dead dragonfly in his hand.

"May I see it?" I ask to be friendly, holding out my hand.

He gasps and pulls his hand away. "How unfeminine," he says, which makes me laugh.

Among the other guests I chat with a man who has recently returned from a year's stay in Toronto.

"Toronto!" I exclaim. "That's where I live!" Amazing the emotion that a quite ordinary home city can arouse in someone who is many months and countless miles away from it. "Did you have an apartment?"

"On Bedford Road," he replies. "Near the university."

"That's the street I lived on!" I almost shout. We mention our various friends, none known to the other, but we are both overcome with nostalgia anyway. He kisses me on the mouth when we part and I kiss him back (although I make sure that night that my room door is locked).

After dinner, chanting groups of Swazi men wearing kilt-like skirts run through the dark streets carrying large branches of leaves, celebrating Dingaan's Day. Some flourish spears (called assegais) and shields made of animal hide, or have horse tails strung around their neck or leopard skins around their waist. One leader sports a magnificent headdress of ostrich feathers. Their chanting is so menacing and their behaviour so aggressive that I can't imagine them being defeated by any native enemy.

"Unlike Swaziland, South Africa celebrates Dingaan's Day to mark the *fall* of Dingaan, not his might," explains Mr. Matthew dryly.

"Who was Dingaan?"

"He was a mighty Zulu leader in Natal who fought the Boers in the mid-1800s. The Zulus were so fierce that when they sent an *impi* out, a fighting unit, to destroy an enemy, the members were massacred if they returned having failed in their mission. Sort of like Stalin killing Soviet soldiers who were captured by the Germans and sent back to Russia after the war. Apparently the Zulus who came north to Swaziland were unsuccessful as fighters, so they were afraid to return home. Instead, they gave a strip of land south of them to the Boers so they'd be protected from their Zulu kin. They settled in what is now Swaziland, completely surrounded by South Africa, and became Swazis."

"Why does South Africa commemorate Dingaan at all?" I ask.

"It's the Nationalists' idea," Mr. Matthew answers. "Dingaan's Day is a South African holiday meant to exalt Afrikaners and belittle natives. The Swazis have taken it over and given it a positive spin."

"Good for them. Do the natives in South Africa celebrate too?" It seems a two-faced sort of festival which, depending on your perspective and nation, can honour both a courageous leader and that person's destruction.

"Heavens no," he declares in surprise. "If the natives there tried anything like this they'd be arrested in no time for disturbing the peace." It is good

Swazis by the roadside in Swaziland.

to report that the majority blacks in a democratic South Africa reclaimed Dingaan from the Afrikaners in 1994, making December 16 the National Day of Reconciliation, which continues to be celebrated each year.

However, I'm glad I can't see into the future. By 2005, Swaziland has been free of British rule since 1968, but it is no longer a relative sanctuary for blacks. An autocratic king keeps the one million Swazi people in poverty, most men are forced to find work outside the country where they come in contact with infected prostitutes, and many women are becoming sex workers themselves so they and their children can survive. Now 39 percent of the adults have HIV/AIDS, the highest infection rate in the world, and sixty thousand children are already orphaned. Some infected men see AIDS as a sign of masculinity and a way to impress their friends; they figure they will die soon anyway, so don't worry about having unprotected sex.

Christmas celebrations as I've known them are muted at Fleur de Lys. Even when the radio is on there are few carols played, and with the temperature in the nineties, who would want to sit down to a hot turkey dinner? However, there is excitement among the natives. The most steady workers, sixty of them, are given a complete set of new overalls. The native schoolmaster and I make up 120 bags, each containing one pound of sweets, and pile them beside the 120 bags holding five pounds of sugar, also to be handed out. The workers have five days of holiday, but since they aren't paid for days they don't work, this doesn't mean much.

The schoolmaster is in charge of Sports Day, which is held three days before Christmas, an event almost identical to the annual Sunday school picnics held by churches each spring on Centre Island in Toronto. About two hundred natives attend, many of them from neighbouring farms which don't hold any festivities. The schoolmaster organizes the races which the young boys, including Enoch, especially love—sack races, foot races, three-legged races and a tug-of-war. The 100-yard dash is so successful that the boys have a 440-yard race as well.

The girls, including Bella, all in rather ragged dresses, have a 100-yard bottle race in which each runs with a ginger-ale bottle full of water balanced on her head.

"Well done," I say to Bella, who has on a brown dress with a brown sash; she didn't win, but neither she nor any of the other girls dropped her bottle. She ducks her head shyly and smiles. Then the girls have a 220-yard race.

The watching mothers chat in groups, jouncing their babies on their knees, each infant sporting a sunbonnet. The men stand awkwardly along the sidelines, either in shorts and shirts or dressed Western fashion, despite the heat, in suits, straw hats, and dark glasses. They shove against each other occasionally in a friendly way to show they're not self-conscious. Mokkies, like the rest of the rangers, wears a smart khaki uniform complete with shiny leather boots, a smart hat, and a rifle which he is presumably not allowed to shoot. I take his photograph with his daughter, a barefoot teenager draped in what look like colourful large kerchiefs anchored by necklaces, bracelets, and anklets. Her headdress is adorned with two large fluffy feathers.

After the races, the men and older boys play a game of soccer.

"Fleur de Lys used to have a soccer team that played workers from the other farms," Mr. Matthew tells me, "but it died out several years ago. Our male dancing team has disbanded too. Now nobody remembers the steps; they stop in the middle of a dance and argue about what comes next."

The climax of the day is the feast, overseen by the rangers, for which the packhouse man yesterday shot three wildebeest and a zebra. The carcasses have been chopped with axes into hunks which are boiling in an enormous open kettle over a huge fire; if a man is very hungry, he grabs a chunk of meat from the kettle to munch on, even if it's almost raw. To go with the meat, hundreds of loaves of white bread, large containers of mealie meal, a vat of kaffir beer supplied by the natives, and twelve cases of soft drinks are piled on long tables. Mr. Matthew has really gone all out!

Before I go to bed that night, Mr. Matthew returns to the house after a visit to the schoolmaster.

Mokkies in his ranger uniform with his daughter
dressed in her initiation costume.

"What a wonderful party you gave!" I exclaim as he comes into the living room.

He sits down heavily, ignoring my enthusiasm. "It's hopeless," he says in discouragement. "Some boys broke into the schoolhouse and stole food stored there. And they broke some furniture."

"Oh, no!"

"Mokkies is dead drunk, even though he's a Watchtower boy." Watchtower "boys," who swear not to smoke or drink, wear a green felt sash with a silver star to reflect their status. "The other Watchtower boys left the party as soon as the drinking got out of hand, but not Mokkies."

"I'm so sorry," I say, knowing how disappointed he is.

The next day, when Mr. Matthew will be away in Nelspruit, I'm invited by a young woman my age whom I'll call Betty to a nearby farm to play tennis and have dinner. I'm keen for a break after so many months of work. There are eight adults present, all very friendly, several of whom have been to the University of the Witswatersrand in Johannesburg. We take turns playing tennis doubles, our pleasure somewhat diluted because any ball that bounces over the low fence has to be retrieved from among scattered bushes where mambas hang out.

"We killed five mambas around here last week," Betty says cheerfully, intending but not succeeding in reassuring us.

Interest in tennis leagues is so great that it's common for teams to drive 150 miles over dirt roads for a weekend to play an opposing group; Betty had asked me to join them on their trips (I played for my university in Canada), but Mr. Matthew vetoed the idea as a waste of time.

The non-playing guests sit about in deck chairs drinking and entertaining themselves now and then by chatting with Betty's two small nieces, both with dirty faces. If one of them cries and becomes bothersome, a native nanny of about ten years of age, waiting in the background, is called to take the child away.

Before dinner, I go with Betty to the outdoor toilet. As we return, she says "You may think me funny, but I have one habit that my sister and I always like to keep."

"What's that?" I say with interest.

"We both wash our hands after going to the toilet. We got used to it at school."

"I do too," I say, trying not to appear surprised at her revelation. We wash them side by side in a basin beside the well.

The dining room is crowded, with eight adults bunched around the table, two screaming children running behind us, and five dogs lying about on the floor. High piles of papers, clothes, and broken toys fill each corner of the room. Two native girls bring around plates of hash, rice, spaghetti, and potatoes. We discuss first the professional tennis group of Frank Sedgeman, Tony Trabert, and Pancho Gonzales, who in October had played matches at the local club in Nelspruit to which most of them had gone. (I wonder nowadays if these poorly paid players even began to imagine what their tentative efforts would lead to fifty years later, with hundreds of men and women earning millions of dollars with their racquets.)

The conversation then turns to Canada. What is it like?

"Women there work much harder than white women do here," I offer, but no one believes me.

"You have electricity and modern equipment to do your housework," one woman argues. "On the farms here we have to struggle with wood stoves, oil lamps, and no running water." As a guest, I don't like to mention that it is the native cook and houseboy, not she, who largely contend with such inconveniences.

Another guest is a nurse from Europe, hired to work in the Transvaal because South African women don't like making beds and washing things.

"I'm leaving Africa as soon as I earn enough money," she says indignantly.

"There's nothing to do in our spare time because it's not safe to go out after dark. And most South Africans look down on nurses because they say we do natives' work. What can we do about that?" No one has an answer.

I drive back to Fleur de Lys well after dark. Mr. Matthew, returned from his trip, is reading in his pyjamas in bed. "I was worried about you getting back safely," he calls out to me as I walk down the hall. "Come and tell me how the day went."

I move toward his bed but stop before I reach his outstretched hand. "I had a good time," I say, not too enthusiastically in case I hurt his feelings. I tell him about the entire evening, and about the general squalor surrounding my hosts' farm, although this feels too much like gossip.

"Europeans talk about how they raise the living standards of natives when they live among them, but often it's the other way around," he says. "Many poor whites sink to living like natives."

"Too bad," I reply.

"Come here," he says again, holding out his hand. I want to put my hand in his but I don't. I stand there silent, not knowing what to do. Looking back, I wonder at my behaviour. How could I not have taken his hand to show my huge gratitude to him for all he had done for me? What harm would that have done? But would he have held on and not wanted to let me go? I'm sure not, but I must not have been sure at the time. Even now, long after his death, I'm not sure. Was my behaviour ungrateful or prudent? Whatever it was, it didn't turn Mr. Matthew against me.

On Christmas Day, Mr. Matthew and I drive to visit his friends, the McIndoes, who enjoy a peaceful life at Bushbuck Ridge on the road to Nelspruit; Mr. McIndoe is one of the three directors of Fleur de Lys. We have a lovely, cold dinner along with about twenty other guests, including seven children dashing about. We each receive a piece of cake for dessert with a prize embedded in it. When I pull out a ring, there's endless joshing about this foretelling of my early marriage, which annoys me. I wish everyone would see me as a zoologist, not a young woman eager to catch a man. But does this mean Ian will wait for me, I secretly wonder. What a pity that young women at the time felt virtually forced by society to choose between having an independent career and getting married. How wonderful that such a choice is often unnecessary nowadays!

I think sadly of the Christmas festivities that my family will be celebrating in Toronto without me. My mother, who as a Dean of Women lives at the Women's Union, has appropriated the whole building during the holidays to house my brothers and sister and their children. They will be having a magnificent reunion, with snow as well.

I'm sad, almost in tears, to be leaving Fleur de Lys, even though I'll be coming back in the spring (their fall), thanks to Mr. Matthew's generosity. I go around to all the people I know to wish them Seasons Greetings and goodbye. First the Europeans, as they like to call themselves—Farnie, Kruger the packhouse man, Van Vechmar the cattleman, and George, who gives me a kiss on the lips.

"Mr. Matthew would kill me if he saw me do that," he laughs.

I laugh too, thinking he may be right. Poor George. After I've returned to Canada he writes me in excitement that a "girl" is coming from Germany to see if she wants to marry him. She later goes back to Germany alone, unsatisfied either with George or with the isolated life at Fleur de Lys.

Before I leave, I ask Mokkies again about any giraffe information he may have collected, but he only chuckles. I don't feel strong enough to press the matter further; after all, I'm not paying him to collect such information, and if he really believes I am working for the racist government, as Mr. Matthew has told him, I can't blame him for dragging his heels. Maybe he'll have something when I come back? Jackson wishes me well, as I think do Enoch and Bella. I shake hands with Watch and Nelson, who look sad to know I'm going.

Cheer up, I tell myself. I'm off again on high adventure to see even more giraffe.

# 10

## Dar es Salaam

Just after Christmas, on very short notice, I'm offered passage on the ship *Kenya Castle,* sailing up the east coast of Africa from Durban, South Africa, to Dar es Salaam, the capital of Tanganyika (now Tanzania). Mr. Matthew drives me to Nelspruit, where I board an evening train for Johannesburg. Each train compartment is built with berths to hold either four or six people, so I'm lucky to have only one other woman for company. On my bunk I open the letter Mr. Matthew has given me at the station; out drops his picture, £15 he thinks I may need on my travels, and a letter in which he thanks me for my "cheerful companionship" during the past months and looks forward to my return to Fleur de Lys.

There is no air conditioning on the train, so my companion and I keep the window open despite the ash and soot that rain down on us as we go through tunnels. In Johannesburg the next morning, I'm met by Mr. Godet, a Frenchman and the third director with Mr. McIndoe and Mr. Whittingstall of Fleur de Lys Citrus Estates; I gather that he is the major shareholder, given the French name of the company. He drives me around the city so I can view the huge piles of mine slag glinting gold in the sunlight, and the native location of Sophiatown.

"The houses are small, but they look nicer than I imagined from Father Huddleston's book," I tell him.

"Father Huddleston's book was most unfair," he says with anger, an emotion shared by most South Africans who have read it. "He wrote nothing about the Roman Catholic Church raising a million pounds to educate natives after the government stopped subsiding their schools."

"How does the government decide who's a native and who isn't?" I ask when we pass a slum where some residents look white and others brown.

"A person is coloured if he's deemed to be coloured," Mr. Godet replies. "This definition gives a little humanity to an inhuman law." It seems to me an odd measure of humanity, allowing a certain amount of leeway in deciding if a person is coloured and therefore worthy of more benefits than if he or she were deemed to be native.

We stop for tea at the posh Transvaal Automobile Club, then go for lunch to Mr. Godet's flat, where his son Paul has prepared a sophisticated lunch, perhaps French, featuring cantaloupe stuffed with raw bacon.

In the afternoon I catch an overnight electric train for Durban which should, in theory, be more comfortable than the smoky steam-engine train but, in fact, is not because it runs on such a narrow-gauge track that it sways wildly, making passengers feel mildly seasick. Early the next morning we cruise through Zululand, where spectacular clusters of green hills and green valleys surround tiny terraced fields; the houses belong to Africans, or more often Indians, each family with a myriad of ragged children.

I spend the day in the steamy heat of Durban, wandering around the streets and passing the nearly empty, dusty Kings Park where, in 1994, a quarter of a million jubilant Africans will gather in a pre-election rally before choosing Mandela as their new president. The next day, New Year's Eve, I board the *Kenya Castle* to be greeted with a telegram from Mr. Matthew wishing me *bon voyage* and good luck.

After four months of monastic farm existence, social life on the ship is euphoric. The steward seats me at a dining table with Martin English, a South African salesman for Lever Brothers products, who has emigrated to Nyasaland; Brenda Vickery from Britain, who is heading to Gilgil to teach; and the ship's doctor, an Irishman who tells wonderful stories and buys bottles of wine for the table. At the New Year's Eve dance to records, Brenda and I pick up a number of partners, one of whom (a Scotsman in kilts) tries unsuccessfully to manoeuvre me into a small upstairs room for sex. I hide in the washroom as midnight looms so I won't have to kiss men as the clock strikes in the New Year.

One afternoon the ship docks at Beira, in Mozambique. Dwarfing small shops made of tin is the magnificent new Grand Hotel, built to kick-start tourism in the country. Smart African waiters in white show us around and serve us small sandwiches and cakes. I have photographs of Martin, Brenda, and me sipping tea in solitary splendour on its broad patio near the blue

Martin English and Brenda Vickery at the new luxury
Grand Hotel in Beira, Mozambique.

Indian Ocean. If I had imagined the future, I would have envisioned the hotel full of wealthy visitors, both black and white, betting money in the casino which will somehow further increase the living standard of the country. In reality, forty-five years later, Paul Theroux reports that whole families now squat in each guest room, with tents rigged on the broken balconies to provide privacy beside smoky cooking fires. Laundry hangs limply on strung-up lines and women empty toilet buckets over the rusting rails onto the patio where we had luxuriated.

During other afternoons I often chat with Martin, who is learning to speak Gujarati so he can sell more soap to the Indian store owners of Nyasaland. I like him because he tends to use big words when small ones would do equally well.

"If your peregrinations take you to Nyasaland, let me know and I can ferry you about," he offers, giving me his address. I'll later take him up on this proposal.

In the evenings on board ship when we aren't dancing, Brenda and I visit the doctor's room to drink and discuss politics. I drink less than the others but, one night, consume four gins which make me see double: the only time before or since I've ever had so many. The doctor tells stories that astonish me. One is about a "high yaller" girl who was so attractive that when she entered a room, the crotches of all the men began to bulge. I know I'm out of my league here.

"Would you like to see my surgery?" the doctor asks me after lunch on our last day before reaching Dar es Salaam.

"Sure," I say. It's a large bright room with two beds in the middle for patients and comfortable chairs along the walls. The doctor shows me these furnishings, then takes off his jacket and trousers, folds them on a chair and lies down on top of one of the beds in his underwear. He beckons me to join him, but I sit down on one of the chairs instead. This doesn't bother him.

"I'm going to have a nap," he says. "You can sleep on the other bed." He immediately closes his eyes and drifts off.

I climb onto the other bed but don't feel the least bit tired. Indeed, the sight of us each lying on a bed in the august surgery makes me laugh. I fetch a magazine to read instead. When I'm bored with that, I go over the Swahili words I'm trying to memorize: *twiga* (giraffe), *simba* (lion), *mingi* (many); I've given up trying to learn the language properly because it has a complicated grammar I know I'll never master.

After an hour the doctor wakes up. "That was good," he says as he gets dressed again. "Now let's go and have tea." As he holds open the surgery door so I can pass through ahead of him, two crew members standing in the corridor regard us with interest but he ignores them. I guess they're amused at him having lengthy medical business to attend to with young female passengers. After tea I don't see the doctor again before we anchor in Dar es Salaam harbour the next morning, but I'll meet him later.

Of all the cities and towns I've visited, Dar es Salaam, the name meaning "Haven of Peace," is by far the most beautiful. My first glimpse of the town is early in the day. Except for the towering cathedral, which stands out sharply against the blue sky, the gleaming white buildings on the waterfront are partly hidden by regal palm trees fringing the harbour's edge. The blue water of the harbour itself is scattered with boats—big ones like us, lighters lumbering toward and away from shore, and flighty white dinghies. Sombre Arab *dhows* cluster together away from the sparkling town as if in disregard of its splendour.

John and Beverley Cairns, to whom I had written from South Africa, are the first visitors to board the ship. It's been over six months since I left Canada, so I greet these compatriots as joyfully as the steward on the *Arundel Castle* greeted me when he found out I was also a Canadian. John, besides being a District Officer who helps the District Commissioner govern the Africans in Dar, is writing a book about his work in his spare time, which will be published as *Bush and Boma* in 1959. Beverley, as well as looking after Sandra, their year-old daughter, and Sandra's native nursemaid, runs a thriving business in native carvings. She spends time bargaining with the best

African carvers for their wares, wrapping the attractive people and animals in paper, surrounding them with sawdust in wooden boxes, and shipping them to exclusive shops in Montreal, Toronto, and New York. In some months she makes almost as much money doing this as John earns as a civil servant. She kindly sells some to me for their cost price, a few dollars each, which I ship back to Canada for presents. (One of them, a male drummer with odd overlapping skirts, I later donate to the Writers' Union of Canada as part of an auction to raise money. Everyone is delighted when reporter Trent Frayne buys it for $300.)

I ask the Cairnses about their activities as we chug ashore in a large, unpainted motorboat but later, as John drives us through the streets of Dar in his station wagon, I concentrate on staring about me. I can't imagine a more exotic metropolis. There is no real industry so the streets and buildings are sparkling clean; no air-conditioning, so barred windows are open to the breezes; and no traffic lights, reflecting the small number of cars.

The airy, white, verandahed dwellings shaded by feathery trees are exotic, but the inhabitants are far more so. Even the Europeans look strange, the most dignified civil servant or businessman dressed fastidiously in white shorts, white shirt, and white knee socks. A few Indians and Africans wear the same tropical uniform, but the Arabs sport white caps and long white cotton robes like voluminous dressing gowns. Some of the Muslim women are in purdah, enveloped in black cloth so completely that only the colour of their bare feet discloses if they are Arab (light brown) or African (dark brown). Other Muslim women follow religious practice less assiduously with only a large cotton kerchief over the head, its edges held to the nose if the wearer fancies some man is watching her; as I watch them, it seems to my mind that this casual arrangement increases the interest and efficacy of flirting. Older Indian women are wrapped in lovely saris, while the younger ones often wear cotton frocks.

To my amazement, one tall, thin African man stalks along the road clad only in a single rust-coloured blanket, with huge wire earrings hanging from the upper edge of his ears. I can hardly believe my eyes—an African exactly as I have imagined he should be.

"Look, look," I cry to the Cairnses in case they have missed this exotic sight. "Look at the African in a blanket! Look at his earrings!"

The ayah, an African nursemaid who is sitting in the back seat beside me, is most amused at my excitement. She can't help laughing when she looks at the man.

"He's just a bush African," she says in Swahili to the Cairnses, who translate her remarks for me. "Imagine being interested in a bush man, and one who

John and Beverley Cairns at ruins of          Sandra Cairns and her ayah.
slave buildings in Bagamoyo, Tanganyika.

doesn't even know how to dress properly!" She continues to chuckle to herself for the rest of the drive. "*Mzungu mjinga,*" she mutters. "Crazy European."

The Cairnses leave me at the small, second-floor flat they have reserved
for me from the Women's Service League. It overlooks the Indian Ocean and
the harbour mouth and is highly suitable with a cold shower, a toilet, a tiny
kitchen with refrigerator (in which I keep my blazer, to protect it from
mildew), a stove, cooking and eating utensils, a small table, three stuffed
chairs, and a single bed complete with a wooden canopy and mosquito netting. There is no glass in the lattice windows or door because there is no cold
weather, and rain is kept out by a roof overhang.

I ask the manager about laundry.

"The houseboy will wash and iron all your clothes," she says.

"Fine. Is laundry included in the rent?" I ask hopefully. My rent without
food (but with a man to make my bed) costs nearly half of what I am to earn,
so this point interests me.

"Oh, no," she says. "Skirts and dresses each cost fifty cents. You can pay
the boy through your bill."

"That's terribly expensive," I exclaim. "Maybe I'll do my own washing."

The manageress looks at me coolly before leaving. There's no doubt
what she thinks about cheapskates. Really, I think to myself. Fifty cents for a
clean skirt! That's far more expensive than in Canada. When I shop for food

in the afternoon, I blush to remember what I had said. Fifty cents turns out to be half a shilling, or seven cents in Canadian money. Even I would question the parsimony of someone who complained at paying seven cents to have a dress washed and ironed.

I plan to work for several months in Dar es Salaam to earn money to finance the rest of my year researching giraffe in Africa. I think of this period as an interlude from active field work, a time to unwind from the steady grind I had imposed upon myself at Fleur de Lys, a chance to write up my giraffe notes and decide what else I should find out about my favourite animal. And, of course, a chance to see how a British colony such as Tanganyika functions, compared to South Africa.

Soon after my arrival at Dar es Salaam, I find a government job as a Confidential Registry Clerk, a classification apparently restricted to white women, and pass a security check which involves swearing that I am not a communist and agreeing that I will never allow an African to handle a government file. I learn to use the flame of a candle to drip red sealing wax on the flap of an envelope before pressing a government stamp into the hot wax to seal the letter. I'm instructed how to operate an incinerator to destroy documents no longer required.

The Department of Labour that hires me doesn't mind that I can barely type, so I have to be content with my wages of £39 a month. Although I'm earning less than I would wish, when there's work to be done at the Ministry it's not exacting—perhaps a letter or two, or a short report to be typed. When there's nothing else to do I type out the giraffe notes I've written in longhand in my room the evening before. Every time our boss from England, Bill, enters the office I tense, afraid he will come over to my desk and read over my shoulder not some minutiae of labour policy but information on "Pawing the Ground" or "Chewing Cud." I've started to read *Crime and Punishment*, and empathize with how nervous Raskolnikov must have felt when questioned about his possible connection with the murders he has committed. However, Bill doesn't work very hard either and he never bothers me.

The other woman in the office is Julie, who came from England on a three-year contract as a secretary and who lives in a large compound for British personnel several miles from our office. She and her roommate, Beth, bought a second-hand car to ferry them back and forth to work, which they take turns driving on alternate days. When Julie drives, she picks me up on the way past my apartment but when Beth drives, she sees no reason to stop for Julie's friend and carries right on.

"Sorry about that," Julie often exclaims when I arrive at work some time after her. "It's nothing personal. Just Beth being Beth." It's wonderful to have a friend I can *really* talk to. For months I've been surrounded by men; I can discuss with them what they're up to and what I'm discovering about giraffe, but not what I'm feeling and thinking about other things.

During January and February my days fall into a pattern. Before work begins at 7:30 a.m., I take a cold shower to wash away the oppressive heat of the night, eat a breakfast of cereal and toast, and swallow a Paludrine tablet to ward off malaria. Then I leave the flat to walk the ten blocks to work. The sight of a European walking is anathema to other Europeans driving downtown (except Beth)—it isn't done. Almost always a car stops to offer me a lift as surely as none offers rides to Africans on foot, even if they're carrying heavy bundles. I walk (or ride) home for lunch, taking a second shower to cool me off after eating a sandwich. After an hour and a half, I march off anew to the Labour Department, again, because of the heat, immediately leaping into the first car that pauses to give me a lift.

At work there are four African messengers, Ezekiel, Mohamed, Mbara, and Cabbage, whose sole occupation is to wait upon us more highly skilled workers. If a typist wants a tea roll, she asks Mbara to buy it for her and keep a few cents of the change. Mohamed, a Muslim, is sent to purchase cokes; Ezekiel, a Baptist, to buy stamps at the post office; and Cabbage to do anything else. When I mention to Mrs. Nicholson, wife of the Labour Commissioner, that I need more anti-malarial Paludrine pills, ten minutes later a messenger arrives to give them to me; she had asked him to bicycle to the hospital and get them free under her name. Julie asks Cabbage to carry out all her commissions because she loves to shout "Cabbage" and have someone respond. One of the messengers seems to have a greatly enlarged scrotum, some disease we suppose, which makes Julie and me giggle although it isn't at all funny. Now I realize that the unpleasant swelling must have been a case of elephantiasis caused by a filarial parasite.

After work finishes at four o'clock I shop at the native market on my way home, invariably buying a lamb chop, potatoes, cabbage, or Brussel sprouts because the variety of local foods is small and the price of imported goods high. Even Jello, my favourite dessert, costs twice what it does in Canada. The only commodity that is cheaper than in Canada is Coca Cola, because of the local Coke-bottling plant. One bottle costs five cents, while imported soft drinks cost twenty cents each.

For much of the late afternoon and evening I work at home at my table, organizing the data collected from the Fleur de Lys giraffe and writing up their behaviour for my report. I also keep up a steady stream of correspondence with my friends and relatives home in Canada. My University of Toronto classmate, Bristol Foster, has written asking me about conditions in Africa since he and Bob Bateman, a former boyfriend, are soon to begin a round-the-world trip by Land Rover. Bristol will later study giraffe behaviour in Nairobi National Park, influenced I like to think by my research, while Bob will become a world-famous painter of wild animals.

I especially count on letters from Ian telling stories about badminton tournaments (indoor tennis was then unknown) and work, and on one occasion about going to see Elvis Presley, the new singing sensation with the swivelling hips. Ian missed writing for two weeks over Christmas, but now seems back on schedule. He has even added a few kisses, xxxs, at the end of some of his airmail forms, thanks to my urging in an earlier letter: "An x on paper is really very tame, but appreciated. You must agree."

Near dusk, for a break from my work, I leave my flat in bare feet to wade in the ocean across the road. At low tide I splash a quarter of a mile toward India before the water reaches higher than my ankles. The ocean bed is sandy, with coral outcroppings that shelter starfish, black sea urchins like pin cushions, and sea slugs over a foot long.

One evening a white lad saunters up to me as I amble along in the shallow water.

"Do you know you have a profile like Doris Day's?" he asks. A good opening gambit, so we settle down side by side on the sand with our backs against a low cement wall, staring out to sea. We chat in a desultory way—where are you from/why are you here/do you like Dar/how long are you staying. He seems pleasant enough until he leans over and begins licking inside my ear. I sit still for a few seconds of surprise, then pull away from him.

"It's getting dark, so I'll be off," I say scrambling to my feet. I stalk back to my flat, annoyed. Maybe I don't have a profile like Doris Day after all. Maybe he just wanted to smooch.

At high tide I can swim opposite my apartment, and farther along the shore Julie and I twice go snorkelling on a Saturday afternoon after work stops at noon. We see hundreds of different creatures of all shapes, colours, sizes, and patterns. Some fish are striped black and gold; others are dotted red on blue; still others flaunt shades of iridescent turquoise, red, yellow, and green. They dart about in search of food, or to avoid yet more colourful pursuers. None flees at our splashing approach. We stop goggling (as we call it) only when we shiver with cold, this a feat in itself in the warm ocean under the tropical sun.

My swimming or wading companions are usually Indians, splashing and chattering in some Indian language. I love the lack of racial segregation in Dar es Salaam, although before I lived in South Africa I wouldn't have thought about it, one way or the other. One group of Indian men sits in a circle at the ocean edge, obviously discussing weighty matters. What could their topics be? A Muslim man sits cross-legged by himself on a cement wall facing Mecca, his shoes placed neatly together in front of him. He gazes motionless at the ocean until after dark each evening, only his lips moving slightly.

Sometimes Julie and I go shopping in downtown Dar. She tells me about Paddy, a man she likes in her compound who may or may not like her; there's terrific competition among the single white women for the few white bachelors in town. We talk about Paddy a lot.

"I went shopping with Paddy last week," she tells me. "He wanted to buy underpants and while he was holding them up to judge their size, I could see the hole in the front. I was so mortified I almost died."

"How awful," I sympathize. Then we both laugh like crazy.

Sometimes I babysit for the Cairnses, or am invited to their place for Sunday dinner. Beverley is a marvellous cook who takes great pains with each course—when I enter the house through the back door, I see soiled pots and pans piled high in the kitchen sink which attest to this. After the meal, the long-suffering maid, who has eaten in the pantry, tackles the dirty dishes, washing, drying, putting everything away, and tidying up.

If there are other guests at the Sunday meal, we retire to the comfortable lounge after eating to chat, surrounded by books and records. Here I meet socially the Labour Commissioner and his wife, the Nicholsons.

"I know men working with wildlife in Arusha," Mrs. Nicholson tells me. "I'll write letters of introduction for you to Mr. Swynnerton, he's the Chief Game Warden of Tanganyika, and Mr. Thomas. They'll be glad to help you study giraffe." She kindly does so but, as it will turn out, her optimism is misplaced.

Unlike the Cairnses, some Europeans have little interest in Tanganyika or in Africa beyond complaining about the heat and the useless servants.

"Africans only stopped swinging from the trees fifty years ago," a smug man tells me. "What could I learn from an ape like that? I have hundreds of years of civilization behind me."

This seems an amazing thing to say, especially since the British in Tanganyika aren't the cream of the crop. Only one officer in the whole Department of Labour has been to university; even the Labour Commissioner doesn't have a degree, which means that I am by far the most educated person in the department, despite my lowly position. These officers are trying to keep a lid on eight million people who are becoming increasingly aware of being exploited.

In hindsight, I realize that I was in Tanganyika at a pivotal moment in its history. The Africans were sick of being bossed about by whites, of being treated as inferior in their own country, and of being paid a pittance for their labour. Although I didn't grasp this at the time, there was a longing for freedom in the air, travelling by drum from village to village, spreading like brush fire to every *shamba*. The Gold Coast would throw off British rule and become independent Ghana during my stay in East Africa. Soon Nigeria would follow suit. Both these countries, like Tanganyika, had few white settlers to complicate the departure of British colonial officers should the "natives" become too unruly to be worth the effort of suppressing.

By virtue of working at the Department of Labour, I'm able to read the confidential files dealing with the "African problem." One of the African leaders in 1957 is Julius Nyerere, who has recently returned from a speaking tour in the United States and who will soon become the first president of Tanzania. A person who I call in my personal letters "a communistic union leader negress from New York" (no mention by me of pride that a woman is so highly regarded in union politics) has followed him to Tanganyika, where she is consulting with the African leaders.

On the several occasions when Nyerere comes to talk to the Labour Commissioner, he sits waiting patiently and too long on a bench in our office. (I don't know at the time that, with a master's degree from the University of Edinburgh, he is actually much better educated than the commissioner; he has even translated the play *Julius Caesar* into Swahili.) Julie and I, who have both read about him in the files, try to observe him, a neatly dressed slim man in his thirties, without staring, as we attend to our typing.

Nyerere's modest demeanour contrasts starkly with his speeches to huge political gatherings, which are transcribed in the labour files. In one he paraphrases Mark Antony's oration using the refrain "They call us Black Nationalists" instead of "For Brutus is an honourable man." He says things like "We only ask when the government will give us self-government, but they will not answer and call us Black Nationalists. Uganda has thirty out of sixty Africans on their Legislative Council and when we ask why Tanganyika can't be like that too, they call us Black Nationalists."

Many of Nyerere's political comments are more personal. He says in one speech, noting that green political uniforms are banned by the government, that people will wear what they like when there is self-government and that the governor would probably be happy if everyone went naked. He admits it is too bad that a European's car was stopped by Africans, but asks how many Africans have been sworn at or beaten by Europeans. When nearly everyone at the rally raises their hand he says that he isn't surprised, because he is often told to "get out" and called a "bastard," even though this word reflects on his mother's character, not his.

John Cairns, who knew Nyerere quite well, tells me in 2004 that these political gatherings were tricky for him and his peers. If they refused Nyerere's request to allow a political rally in Dar, fearing that some Africans might become drunk and violent, telegrams of complaint would immediately be winging to Britain which was responsible for Tanganyika as a Trust Territory. He feels that, in general, the British colonial officers were very fair to the Africans.

The 1957 files show that in the past few years, eighteen thousand Africans have formed twenty-eight unions of sorts, each hopeful of change but handicapped by lack of money, illiterate members, and leaders with little education. A year earlier, in 1956, these unions joined to form the Federation of Trade Unions; it has a post office box number but not a telegraphic address which, at a cost of £2, is too expensive. The related Tanganyika African National Union (TANU) is also fighting for independence for the country.

The files describe how in October of 1956, when four waiters were dismissed from their hostel for being drunk while serving at a banquet in honour of Princess Margaret (an unforgivable sin to the British, given the occasion), Africans throughout the country were called on to strike. By December, just weeks before I arrived, Dar itself was more or less shut down; the British employers were at a loss because their servants and workers hadn't made any demands to which they could respond. Rather, TANU leaders broadcast that all workers must receive at least a 200 percent increase in wages, which seemed to the British completely unreasonable. The walkout soon collapsed because the strikers received no strike pay and could no longer afford to carry on. As a result of the strike, no workers made more money and many lost their jobs.

When I begin work at the Department of Labour in January 1957, TANU is organizing another country-wide strike of workers for February 11, which will involve a shutdown of all essential services. In response, the government is working out plans for a State of Emergency with martial law to be enforced, which should be exciting if not scary. Trade Union officials are flying to Dar from Britain to help cope with this job action.

"We'll have to cook our own meals!" says Julie, pretending to be horrified.

"And make our own beds and do our own laundry," I laugh.

None of these dire possibilities occurs, however, because TANU suddenly calls off the strike; it's too difficult to organize it well enough to be effective in a short time.

"That's good," says Bill sarcastically. "I've heard that workers are being beaten up if they're afraid to strike, so now they'll be safe."

Official comments that accompany political speeches in the files indicate that the government would like to ban Nyerere, but realize this would look undemocratic and give TANU wonderful propaganda. The government has sent a few Africans off to study trade unions in Europe and in Japan—after all, its ostensible aim is to prepare Tanganyikans for future self-government—but it is worried that when these men come back they will direct their new talents toward political upheaval. One Tanganyikan newly returned home reports meeting a policeman in Hamilton, Ontario, who tells him that trade unions should not have to ask the government if they can hold a meeting. In Canada, the policeman says, trade unionists are citizens and therefore are, in effect, the government.

Even mild signs of a need for change in Tanganyika are considered heretical by Europeans; according to the files, one native Department of Labour employee complained in 1953 that his wife was not allowed to have a baby in the European hospital even though he paid the same tax as Europeans. This complaint would seem, logically, to be justified. However, the woman gave birth to her baby elsewhere and the man is still considered dangerous in 1957 because of such ideas; all his activities are carefully watched by the British officers.

Tanganyika wins its freedom in 1961, only four years later. Visiting Dar a few months after this, journalist Ryszard Kapuscinski describes meeting a number of colonial administrators who had entered their office in the days after Independence to find a smiling local African sitting at their desk. The process of Africanization began immediately, with Africans taking over the positions of whites, and often their cars and homes as well. In the next weeks and months Africans became used to the luxury of servants, a swimming pool, and a lovely garden while the Brits packed up and retreated back to a wearisome life in cold Britain. With so much at stake for African leaders, a fierce struggle for political power and wealth began and continues to this day in most countries of the continent, leaving most of the populations in poverty if not open warfare.

Julius Nyerere, as president of the one-party TANU government, struggled to bring prosperity to Tanzania, the country established in 1964 with the inclusion of Zanzibar. When he realized that members of parliament were out of touch with the common folk, he made them walk the two hundred dusty miles to Mwanza for the annual party meeting. On the way they slept among villagers and ate from their common pot. He tried to create a socialist nation with the help of China, based on tribal values, but this proved impossible: the combination of exploitation by first world powers, tribal rivalries, colonial legacies, and scarce economic resources was too much. Tanzania remains one of the poorest countries in the world, with tobacco and sisal plantations abandoned while Tanzanians live hand-to-mouth, their government without finances to improve conditions.

In 1957, the irony of life in Tanganyika is that although the Africans rightly see the British among them as rich beyond belief, the whites see themselves as hard done by. They do have a house, a car, and servants, but they are often in debt to pay for these. Most of the married women in Dar es Salaam must work to send their children away to school, since those older than seven are educated at boarding schools where the climate is healthier, either up-country or in England. Civil servants dream of their six-month holiday in Britain as a welcome respite after each three-year tour of duty in the torrid colony.

Rereading my old letters from 1957, I would like to discover that I was sympathetic to the Africans, that I realized how difficult it was for them to make a better life for themselves when they had little money and no power. Why did I idealize blacks in the abstract but not give them credit for wanting to be independent of British rule? For having ambitions similar to those of whites? Instead, I find to my horror that I was brainwashed by the British. I wrote then that the wage demands of the labour movement were "so farcical that everyone laughed." They (the British among whom I lived) may have laughed, but it surely wasn't funny for workers to earn a wage that barely allowed them to live above poverty. I wrote sarcastically that their propaganda slogan "God Is with the Workers" probably meant "they expect help from there," as it wasn't coming from anywhere else.

When I return home to Canada and give my brother Hugh an analysis of the African situation as I see it, he bursts out laughing.

"That's ridiculous," he says. "How can you believe the British are really preparing the Africans for self-government? How many of them are in schools? How many in university? How many doctors are being trained? How many lawyers? Are they given courses in politics? In self government?"

On that day, in a flash, I realize he's right. How could I have been so deluded? I've kept reason at bay in one part of my brain—that the British are idealistic and in Africa only to help the natives—just as I have accused the whites of doing in downgrading all blacks. It's a sobering thought. I always sympathized with the natives while in Africa, but I didn't appreciate that they wanted and needed freedom from oppression just as I would. The only good thing about my belated realization is that, in the future, I will try actively to help those who suffer from discrimination, especially when I become a feminist activist aware of pervasive sexual as well as racial discrimination.

One afternoon, sweaty as ever from the heat, I'm ambling home from work (it's Beth's day to drive) when I glance over at the harbour and see the *Kenya Castle* at anchor. Immediately I perk up. I turn on my heel and march to the docking area where I board a launch carrying supplies to the ship. It seems far too long a time since I sat around laughing and drinking and enjoying myself with a group as we did a month ago on this ship. Once aboard, I ask for the doctor, who is delighted to see me. He arranges for me to stay to dinner, which is delicious. We have a fine time laughing and joking.

"I'll walk you home," the doctor offers when it's time for me to go. This is kind of him because it's been dark for hours and the streets are poorly lit. We take a lighter to shore, then stroll slowly along, enjoying the balmy evening.

"Thank you for the lovely dinner," I say when we reach my flat, shaking his hand.

"Can I come in for coffee?" he asks. "It's a long walk back."

"Of course." We're still enjoying our conversation.

After coffee he suddenly says, "It's too late for me to get back to the ship tonight; the ferry service will have stopped."

"You can probably get a hotel room," I say, startled. I feel a little guilty that I've kept him so late.

"No, I'll stay here," he decides. "We sail early in the morning, so I don't want to go chasing after a hotel room at this hour."

"But I've only got one bed," I object.

"There's room on it for two," he says calmly, although it's a single bed. "The mosquito netting will keep us from rolling out."

"But I can't go to bed with a man," I protest, feeling desperate. How can I get rid of him?

"Don't worry," he remarks. "I won't touch you if you don't want to be touched." Without saying anything more he takes off his trousers, shirt, and socks, folds them on my chair and climbs into my bed in his underwear.

I don't know what to do. I look at him crossly but he has his eyes shut, his mouth set in a silly smirk. I finally change into shorts from my skirt and lie down on the floor. I'm angry that I have to sleep on the floor in my own place. But I can't sleep. The floor is hard, mosquitoes keen around my head (do they carry malaria?), and I can feel cockroaches and ants running up and down my arms and legs. Finally I can bear it no longer. I stand up, pull out the mosquito netting from under the mattress, and climb into bed beside the doctor.

The doctor doesn't say anything, but after a few minutes he leans over to kiss me. I push him away roughly. He shrugs, turns over with his back to me, and falls asleep almost at once. I doze rather than sleep for the rest of the night, unable to move because there is so little room on the bed.

In the morning, the doctor is in the best of spirits so I cheer up too.

"Didn't we have fun last night?" he insists.

"You have no idea how awful the bugs were on the floor," I laugh.

We skip breakfast because the doctor is in a hurry, but after we've left the flat he stops suddenly.

"I left a French letter, a condom, under the pillow," he says good naturedly, "but never mind, you can keep it. There isn't time to go back."

I'm horrified. "There's a man comes to make my bed each day. He'll find it for sure!" The doctor leans against a palm tree as I rush back to my room, find the small packet, and stuff it into my purse. As I reach the sidewalk again, one of the secretaries stops to give me a lift. I'm not sure what to do. Should I hop in and leave the doctor behind? Should I make an excuse and refuse the lift? Before I can make up my mind, the doctor walks to the car and introduces himself.

"He's the doctor from the ship in the harbour," I explain weakly as we both climb into her car. "He came up early this morning to say hi." Even I know this sounds ridiculous.

The secretary gives me a smirk and chats with the doctor as she drives into town. I never see the doctor again, but several months later I notice his picture in a Durban newspaper. He is marrying the daughter of a wealthy businessman in a huge wedding. He must have been engaged to her when he slept in my bed.

All morning I worry about the French letter in my purse. What if someone sees it? What kind of girl will they think I am? When I get home from work at lunchtime, I take it out of its paper envelope to see what it looks like. When I unroll it I find that it's huge—really wide across and incredibly long. I can't believe it's so big. The only penis I've ever seen is that of Michelangelo's David, in Florence, which is tiny in comparison (and indeed smaller than normal I read later, for reasons known only to the sculptor). How could anything so large fit into anyone? I'm not into thinking in more personal terms.

In revulsion I rush into the bathroom and flush it down the toilet. To my horror, it bobs up again in the bowl. I try a second and a third time to flush it away, but without success. It still floats flaccidly on the water. Finally I grab a scissors and cut it into small pieces, which eventually flush away.

When I recently shared my account of this adventure with the three other members of a Writers' Workshop to which I've belonged for twenty years, all women, they were dismayed.

"He was almost a rapist!"

"Couldn't you have called the police?" (I had no phone and no belief that police dealt with such situations).

"You were so naïve!"

"We were all so lucky in those days," we agreed. Lucky not to be physically harmed by men.

On one of my daily strolls along the beach I find a brown curlew crouched
helplessly, probably injured by a dog. When I lift it onto its feet it can walk,
but it can't rise by itself from a squatting position. It's sixteen inches high
with long legs and an even longer beak, curved under at the tip. It can't look
after itself so, tucking it snugly under my arm, I carry it to my flat to save it.
It sits morosely in a box of grass I prepare for it, ignoring the tasty beetles and
nuts I offer (what on earth do curlews eat?) and unable to drink from the
bowls of water I set out because of the excessive length of its bill. When I put
it on its feet, it stalks soberly about, peering into the kitchen and snooping in
its box. Otherwise it rests quietly in its corner.

The next day I carry the curlew to work in a paper bag. Bill, typically
British in his solicitous attitude toward animals, insists that it visit the veterinary clinic. As its escort, I'm given the morning off work and am driven there
by a chauffeur in a government limousine. The veterinarian looks surprised
when the curlew and I arrive in such a grand manner. He examines the bird
carefully but can find nothing wrong with it, so we are driven back to work
in the limousine. That afternoon I release the curlew, hoping that it will
somehow forage for itself more capably than I've been able to do for it. Without looking at me, it splashes phlegmatically off into the shallow water.

Besides the curlew, my flat houses a number of indigenous animals. The
welcome ones are two green lizards who patrol the ceiling for insects during
the evenings; at first I hadn't been too keen to share my room with lizards, but
as there is no glass in the windows there is no way to keep them out. Unwel-
come are the hoards of tiny ants that stream perpetually through my flat. One
ceaseless column marches in the front door, up the wall, across to the kitchen,
over the refrigerator and the counter, along the shower by way of the ceiling
and into a crack in the shower wall. This pilgrimage never diminishes in
number during my two-month stay in the apartment; indeed, other bands of
ants also materialize occasionally, most often in the kitchen. If I spill a drop
of Coca Cola, fifty ants arrive to celebrate before I can wipe it away, or realize that there is anything to wipe; when I cut a slice of bread, I wave it in the
air before eating it so that the resident ants can climb down my arm rather than
be consumed.

Sometimes, in sheer rage against the purposefulness and infinite number of my enemy, I massacre with my hand every ant that I can reach, deci-
mating scores with every sweep of my arm along the wall (an activity I would

never indulge in now, as an animal rights activist). But more ants stream in the door and up the wall in a continuous file, following the line of slaughter unswervingly to the hole in the wall of the shower.

I also lose my temper during a flying ant—flying termite?—"attack." As I sit working at my table one evening, several of these creatures with two-inch wing spans circle my head, one landing beside my pencil and crawling over my paper. I leap to my feet to close the windows, then remember I can't do this—there are no panes of glass or shutters. Before I know it, the insects arrive in hordes, sailing in the windows and swooping about the room, bumping indiscriminately into chairs, walls, and me. Why flying ants should choose to gather in force and storm my apartment on one specific evening, I don't know. The noise of their collisions, the annoyance of being struck repeatedly in the face, and the sight of the room swirling with intruders fills me with rage. Picking up a folded newspaper I gallop about the room, striking frantically at all the flying ants I can reach and waving savagely at those I can't. Flying ants drop their wings in time of stress and crawl off along the ground, which the ones I hit in the air promptly do. So, between the airborne attacks, I swat everything that moves on the floor as well.

At the end of an hour, their assault (and mine) ends, leaving wings and crushed corpses littering the floor. (How embarrassing to have to report this when at present I will no longer even kill a fly.) I retire to bed exhausted. As I drift into sleep, an occasional rustle, as a newly arrived flying ant bumps into the mosquito netting around the bed, reminds me that my victory is not complete. What will it make of all its fallen comrades?

Another intruder in my flat is a four-inch, hairy spider who takes up residence on the floor of my shower. Two of its legs are gone, leaving it with six. On its first day at the flat I take a shower with the spider, reasoning that I will stay inside the circle of falling water while the spider will remain on the dry periphery. This theory is sound, but when I get soap in my eyes and can't see, I lose my head. Every drop of water feels like a spider touch. With a wild leap I flee into the kitchen, swiping at myself with the towel in case the spider accompanies me. When I dry my face, I see the spider still cowering away from the water in the shower. I finally capture it under the vegetable colander, transport it to the window and drop it to the ground, where it ambles off toward the garage.

Before I arrived in Dar, I imagined the Cairnses leading an idyllic life with a sense of mission, but with none of the discomforts often attached to a mission—a life spent in helping the less fortunate, but with servants and an interest in Africa as well. Such an existence may have been possible many years ago, but it is so no longer. Already the strikes and political meetings by

Africans have created much unrest. There is talk of the Africans obtaining self-government in fifteen years, a period judged to be ridiculously short by Europeans. Nevertheless many of them, including John and Beverley, are planning to leave. In fact, the nation becomes independent in 1961, only four years later. John goes on to a distinguished career with the World Literacy Program, and later visits and assesses projects around the world that are or might be funded by the Canadian International Development Agency (CIDA). Beverley continues as a brilliant artist and sculptor now based, like John, in Elora, Ontario.

# 11

## *Zanzibar*

Early in March 1957, collecting the money I've earned working for two months at the Department of Labour and mailing off to Griff the report I've completed on the Fleur de Lys giraffe, I set off from Dar es Salaam to continue my exploration of East Africa and its giraffe. I'm anxious to see the *rothschildi, tippelskirchi,* and *reticulata* races that live here, all with distinctive spotting, but before I do this I visit two legendary places I've always dreamed of seeing: Zanzibar and Kilimanjaro. Will I ever be in this part of the world again? I assume no. I plan to take photographs of these wonders to show friends and relatives back in Canada, whom I am sure will never have the opportunity to visit these faraway places. Little do I imagine how the growth of the airplane industry will make even the farthest corners of the world easily accessible during the next few decades.

To reach the island of Zanzibar, fifty miles to the north of Dar es Salaam, I fly in a silver Dakota along with four Europeans, two Africans, and twenty-two Indians: my first flight ever. The Cairnses drive me to the airport where I board the plane sitting on the hot tarmac. The heat is so suffocating inside that I have to steel myself not to run to the door and pound on it to let in some air. It's better once we're airborne where, from my window seat, I can see coral reefs and dark seaweed in the shallow turquoise water. When we arrive at the Zanzibar airport, a single-storey, yellow building, a talkative taximan agrees to drive me the seven miles to town.

"My ancestors came from Muscat," he tells me proudly as we set off.

"Aren't you an African?" I ask, staring at his dark skin and black woolly hair.

"I am certainly *not* an African," he replies testily. "I'm an Arab. My ancestors came from Muscat. Africans are bushpeople. We have no use for them."

We proceed in silence for a few minutes, past forests and small fields.

"I'll drive you on a tour of the island for a pound," he then offers kindly.

"Thank you, but I'll see how it goes," I answer. A pound is a lot of money.

"The island is large," he says, although at fifty miles long it is small.

"I'll see," I repeat.

When we near the hotel, he says "I'll drive you around the island for nothing." This seems like *too* good a bargain.

"I'll see."

The hotel I've booked into on the Cairns's advice is an old Arab mansion with thick, white, plaster walls inlaid with black, square-cut timbers, reminiscent of an Elizabethan dwelling in England. My large room is eighteen feet high, with cold running water and a ceiling fan. The handsome main door of the hotel is like many others in Zanzibar—of elaborately carved, thick, burnished wood, studded with polished brass spikes.

"The spikes protect the houses against elephant attack. The elephants don't like bruising their heads when they charge," a fellow guest jokes to me with a smirk on his face as I admire it.

After lunch I set out by myself to explore the town, to the distress of four Arab guides clustered around this door.

"Only ten shillings to see the sights," one says.

"Five shillings?" a second suggests.

"You'll get lost," a third calls after me. Two follow me for several blocks but then, discouraged by my determined advance, wander back to the hotel to capture easier and wealthier prey.

The streets through which I stroll are narrow and crooked, like those in Venice, with buildings three or four storeys high which nearly block out the sky. Few passageways are wide enough to allow two cars to pass. Some accommodate small cars so snugly that pedestrians are forced into side streets at their approach; the smallest serve only those on foot.

People living in these passages sit cross-legged at the doorways of their tiny shops, nodding in a friendly manner at each other and at me. Some greet me with a friendly "jambo." The men wear traditional white robes with red or white caps on their heads. The women's dress is more varied, with the Sikhs wearing silk trousers, and the Hindus and Muslims in saris or cotton frocks with black *buibuis* or cloaks over their heads, remnants of purdah. I pass one young African girl strolling hand in hand with a beau, her *buibuis* covering her hair but her lips red with lipstick. The man and the lipstick surely negate the purpose of the *buibuis?*

I wander on in a reflective trance, my historical memory very much alive from the reading I have been doing. In the 1860s, David Livingstone and Henry Stanley traversed these very streets while buying and organizing supplies for their expeditions by foot into the interior of Africa. It's only 125 years since slaves were imprisoned in Zanzibar with manacles on their legs and dog collars around their necks, waiting to be sold in city markets. Many of the Africans will have ancestors who lived through this awful time.

I snap out of my reverie as I reach the harbour, which sparkles with new paint—not its habitual condition, I learn, but spruced up for the recent visit of Princess Margaret to the island, which is a British protectorate; the princess doubtless appreciated this because not long before, she had renounced, under pressure, her plans to marry the divorced Capt. Peter Townsend. Zanzibar is apparently fond of all royalty, including its own Sultan, because everyone I talk to praises him excessively. He lives in a huge white palace on the waterfront, his red Daimler parked at the door and his private yacht moored nearby in the small deep-water berth.

Beside the palace stands a huge, pink, formerly Portuguese fort that was built in the seventeenth century but is still in excellent condition, with towers at the four corners. It is now used as a Muslim women's club where, with no men about, woman and girls (many of them Girl Guides) can remove their veils and play ping-pong. For the literate among them there is a small library with books in Arabic, Swahili, Hindi, and English.

Manu Shah, an Indian bank clerk, introduces himself and guides me over the fort. "Zanzibar is a wonderful place," he confides. "There's no racial conflict at all."

"Even with Europeans?" I ask in surprise.

"Three hundred Europeans live here, but there's no discrimination." From the roof he points out the nearby House of Wonders and the new Standard Bank, whose Arabic architecture fits in perfectly with that of the older buildings.

"Why is a Hindu man allowed into a Muslim women's club?" I ask.

Manu laughs. "This is Grand Fête Day," he exclaims, "one of the biggest days in town. Everyone comes to the fort to have fun. All the proceeds of the fair go to the poor, because there's no begging allowed in Zanzibar."

Manu leads me into the huge courtyard of the fort where, sure enough, everyone *does* seem to be present with the exception of the three hundred Europeans; I'm the only white person there. As Manu turns to leave me, we exchange addresses. When I send him a greeting card the next Christmas he replies, wishing me luck and health during the year and asking if I would like anything from his side of the world—perhaps stamps? He doubts that anything I saw in the rest of my time in Africa was as good as Zanzibar—everything there is so green and fresh. He recalls that I came into his bank to cash some travellers' cheques, and tells me that for the first time ever Zanzibar has held elections for Legislative Council Members. He ends his short letter in a practical manner—"There is nothing more to pen so good bye for the time being. Meet you in next letter."

At the fête hundreds of people mill about, talking and laughing. Dozens of little booths, operated by Boy Scouts and Girl Guides, line the basketball court in the middle of the field. They're identical to those I've frequented at fêtes and fairs in Canada, featuring ring tosses, coconut shies, white elephant stands, fish ponds, and raffle sales.

When I've used up all my East African change at the fair without winning anything, I continue my walk along the shore. This time four little Indian girls follow me.

"Good morning," I say to them. They dissolve into giggles; after all, it is afternoon.

"Good afternoon," I try again. This time they laugh even harder, nudging each other at the same time. I walk on, pleased at my easy humour. I pick up a dead crab lying on the path which attracts two Indian boys, who come close to examine it. One points to its eye.

"Eye," he says.

The other points to its mouth. "Ear," he says.

Several old Arab men on a bench peer at the crab curiously, then discuss it in Arabic. Soon a number of African boys who have been swimming rush up to see what we're doing. They're less impressed with my find. They think the crab should be alive to merit such a fuss.

It's now growing dark so I head back toward the hotel, but am soon lost in the winding lanes. A man in a red fez who has seen me doubling back and forth finally asks me if he can help.

"I'm lost," I admit. "I'm trying to find my hotel."

"My name is Seyyid Asswedi," he says, pointing out the way for me to go. "I own a Greek Guest Home where you can stay if you come again."

"Thank you," I reply, both for his directions and his invitation.

"Would you like to drive around the island in my car this evening?"

"No, but thanks," I say. How could I see anything in the dark? He phones the hotel later but I'm having a bath at the time so don't get the message from the main desk. (It seems odd that some Arab men are incredibly friendly while others are anything but. Years later, when on a camel safari in the Sahara, I meet a few holy men who refuse to shake hands because I'm a woman and an infidel, while other Arabs ask me to marry them on a few minutes acquaintance.)

In the evening Jock Boult, a guest at my hotel whom I had met on board the *Kenya Castle,* asks if I want to go to a variety show at the Portuguese fort. "Come with me, since I'm married. I'll protect you from the wolves," he says.

"The wolves?"

"Yes. There are lots of bachelors on the island who don't see a girl for weeks on end. It's best to keep away from them." He must mean European girls, because there are many African and Indian young women about.

"What do the bachelors do with their time then?" I ask.

"Drink," Jock confides darkly. I'm certain I'll be safe with him.

The show is held in the fort's courtyard, where about a thousand people, mostly men, are gathered to watch. The basketball area is now covered with chairs although many hundreds of the audience are also seated on the ground. Two Indian boys are ordered by an official to give up their seats near the stage for us, much to my embarrassment. Jock, however, is quick to sit down and make himself at home.

The entertainment begins with a dance of precise angular movements by a tiny, exquisite Indian girl set to exotic music. Next a tall Indian man, accompanied by an orchestra of seven exotic instruments, sings a song full of strange cadences, the deadpan expression on his face never altering during his rendition. Is he singing about love? Or death? Or pastoral loveliness? Or the moon? There's no way of telling. He receives only scattered applause.

The audience is far more enthusiastic about two African male jazz entertainers up next who sing American love songs with great zest, wiggling their hips in time to the music while the audience shouts in wild approval. Is this too much for Jock?

"We've got to leave now," he says, turning to me when the set is finished. "I have to work in the morning." I'm annoyed that he is so bossy, but don't relish finding my own way back to the hotel in the dark. When we stand up, an Indian couple grabs our seats so the two boys we had displaced are still out of luck. After we've pushed our way through the crowds to the fort entrance, a man takes Jock aside to talk to him for a few minutes. Jock feels forced to introduce me.

"This is Othman Sheriff, a veterinary officer. He says that, if you like, he'll drive you around the island tomorrow; he has business to look after and you can go along for the ride." Jock is frowning. Is this a wolf to be avoided?

"I'd love to go," I say to Othman. "Would I really be no trouble?"

"I'd like you to come," Othman tells me seriously. "I'm always happy to show anyone the wonders of Zanzibar."

"Thank you so much!" I'm thrilled.

"I'll call for you in the morning then, about nine."

Jock stalks in silence beside me back to the hotel.

"Is even Othman a wolf?" I finally ask.

"He's an African," Jock replies coldly.

Othman is fiercely African, I learn the next day, as we drive out of town and I ask him if he is partly Arabic because of his lightish colouring and straight black hair.

"I haven't a drop of Arab blood in my veins," he retorts angrily. "I despise the Arabs. Look at their history as slave traders, bringing misery and death to millions of Africans. Their cruelty is only exceeded by their cowardice."

"Do you like the British better?" I ask hesitantly.

"Yes," he answers soothingly. "Much better. They govern as well as anybody. As long as they stay in Zanzibar the Arabs will not hold power."

In his married life, Othman tells me, he faces a problem common to many educated Africans. He married his wife many years ago at the insistence of his family, but the marriage has never been a success. They have nothing in common. She is in purdah, she knows no English, and the ties of her birth family are so strong that she stays with them in the neighbouring island of Pemba, although Othman lives and works in Zanzibar. Oh, oh. I think of Jock's wolves. This tale sounds suspiciously like a "my wife doesn't understand me" story.

We drive through the island on a paved highway shaded by dense tangles of coconut palms and clove trees which produce the two main crops of the island. African huts are scattered in small clearings featuring banana trees or small rice fields. In one such paddy an old woman crouches, planting rice sprouts one at a time.

We stop at a huge government cattle ranch near the centre of the island, where tsetse bush has been recently cleared so that zebus—tropical cattle with large humps—can be raised. Bushes that favour tsetse flies still cover half of the island, but luckily the species of fly here doesn't spread sleeping sickness to people; it only infects cattle. In a nearby, wooded area we visit a bullock who has been tethered there for a week. Othman collects blood from its ear so he can find out from a blood smear if the animal has been infected by tsetse flies during that time.

Near the bullock, ten lithe Africans dressed only in loin cloths scramble up and down palm trees in a large palm grove. From the tops of the trees, each man throws any ripe coconuts to the ground to be collected later to produce "copra," the name for dried coconut meat. As he does so, he sings out loudly in a high note which a counter, sitting on the ground with a notebook, recognizes and notes down under the man's name.

Othman Sheriff collecting blood from a tethered bullock.

"The men earn about fifteen shillings a day based on the number of trees they've climbed," Othman tells me.

Othman stops at a copra factory, where fires burn continually to dry out the split coconut halves. The brown fuzzy shells are used for fuel while the coconut oil extracted from the white "meat" inside is made into food, soap, hair-oil, and margarine. Othman presents me with half a coconut as a souvenir. I carry it with me for three weeks through Africa, until I realize it's attracted a stream of ants and several cockroaches into my suitcase.

Farther on in our tour we pause at an ocean beach where a thin African man is digging in the sandy soil for long red worms to use as fish bait.

"Could I take his picture?" I ask Othman.

"I'll ask him," he says. The men chat for a minute.

"He wants to know if his picture will be shown in the movie houses in Zanzibar," Othman says.

I shake my head.

"He says all right, then. But he wouldn't want his friends to see him in the movies." The fisherman continues to dig as I photograph him. When I get back into the car, Othman touches my arm, then tries to draw me toward him.

"I get very lonesome," he says.

I pull away and put my hand on the car door handle. If I have to get out to escape Othman's advances, I have no idea how I'll get back to the hotel.

"Sorry," he apologizes, seeing my fear. "I didn't mean to upset you." The danger is past.

On the road back to town, the Sultan's red chauffeur-driven Daimler approaches us, prompting Othman immediately to stop his car as a show of respect. The Sultan waves to us as he passes and throws pennies to some Indian children standing nearby.

While Manu Shah told me that all was sweetness and light in Zanzibar, Othman, who is an appointed member of the Legislative Council, has another perspective. "Recently the Arabs decided to boycott the Council," he tells me as we drive along. "The Sultan asked one particular Arab to attend the next Council meeting, which he did, but he then received threatening letters and was later killed by an assassin. The assassin was caught and sentenced to death, but the Sultan's wife went on a hunger strike until the Sultan changed his sentence to life imprisonment."

As I thank Othman back at my hotel for his kindness, he suggests I visit him in Scotland the following summer where he expects to be taking a course. When I write to him in June, he replies from a cattle breeding centre near Glasgow. I don't manage to meet up with him, but many years later I read an article by another woman tourist to Zanzibar who describes the pleasure of being shown the highlights of the island by one Othman Sheriff.

Yesterday Manu Shah had told me there were no racial problems in Zanzibar. By contrast, Othman indicates that hatred between the Arabs and the Africans is deep-seated, certainly not surprising given the history of Arab slavers rounding up African victims, and using the island of Zanzibar as a holding place before sending them into slavery in many countries of the world. Unfortunately, Othman was right. Six years later, in December 1963, Zanzibar becomes a constitutional monarchy and although black Africans win a majority of votes in the election, Britain supports the Arab minority party, which forms the government. Almost immediately there is a coup, with the toppling of the Sultan and the massacre of large numbers of Arabs who comprised 20 percent of the populace. Obviously I had been wrong in assuming that the Sultan was universally well-liked.

On my third and final day in Zanzibar, I want to photograph an Arab *dhow*, that exotic symbol of another world. As I walk through winding lanes toward the dock area, an Arab boy stops me, pointing first to my camera and then to himself. He wants me to take his photograph. When I have done so, he hands me a letter so I can copy his address to send him the picture.

A few minutes later an African man also wants his photo taken. While I fiddle with the camera he stands as straight as possible, his chest stuck forward importantly. After I've clicked the shutter, he holds out his hand. He wants the picture. Surprised, I shrug sadly and point to the camera. That's the villain. The man looks at me with astonishment, then with resentment. I point to the camera again. Suddenly he bursts into a roar of laughter. He has been fooled. He acts out how he stood so straight for so long, then held out his hand but received nothing. He laughs on and on at this wonderful joke on himself. As I wave goodbye and leave, he's retelling the tale to a friend, first standing rigid, then holding out his hand, then exploding with mirth. If I had known enough Arabic to ask him for his address, I would have been able to send him his picture, too.

Near the docks I'm stopped by an African official dressed in white shorts and shirt. "May I help you?" he asks in English.

"I'm hunting for an Arab *dhow.*"

"Ah, you have come to the right place," he replies without surprise. "I am the *dhow* inspector, Abdul Hamdany. Come with me."

I follow Abdul through a series of narrow streets past hundreds of men lounging on either side of our route. All of them turn to stare at me.

"Men in Zanzibar are lazy," Abdul explains. "Our only sizable exports are cloves and coconuts so the men sit about most of the year, waiting for the trees to produce."

"*Salaam alekum,*" Abdul salutes several friends.

"*Alekum salaam,*" each replies solemnly.

Around us, tiny booths run by Indians sell oddments of cloth, tinned food, and dried shark meat. In the infrequent open squares, small fish, sea slugs, and spaghetti are spread out in the sun to dry.

We pause at a store selling fine Chinese merchandise. "The clerk is Chinese," says Abdul. We stare at the man. "There are only forty Chinese in Zanzibar," he continues. The clerk stares back at us.

Near the water Abdul points out several huge piles of mangrove logs. "They'll be shipped by *dhow* to Arabia and India for use in building," he says. "When the monsoon winds change in six months, the *dhows* return to Zanzibar and the mainland of Africa carrying shark skins, carpets, baskets, and of course roof tiles. Roof tiles break too easily if they're sent by regular shipping lines. The trip by *dhow* takes two months."

"And the crews wait here four months for the winds to change?" I ask in amazement.

"The crews, yes," Abdul replies. "Some captains and owners, no. They fly home to spend the time with their families."

We see the *dhows* now, about fifty listing heavily in the mud, waiting for the tide to come in and straighten them into self-respect. Each boat is about fifty feet long.

"*Dhows* are made entirely of wood," Abdul says, "the way they've been made for centuries. Come and visit my uncle who owns this one here."

An Arab dhow in Zanzibar.

Abdul strides to a *dhow* tied loosely to the dock and climbs ahead of me up a treacherous rope-and-board ladder onto the deck of the boat. A number of Arab men in turbans and skirts or shapeless gowns are hunkered down there, squeezing limes with sticks of wood. The juice is funnelled into old whisky and beer bottles to be sold in Arabia.

Abdul's uncle, the captain, a small turbaned man dressed in a grey gown and with bare feet, comes forward to meet me. Abdul explains my presence in Arabic while the captain and I nod and smile at each other. The captain motions us to the stern of the boat, where a low canvas sheet shades the deck from the sun. I sit as directed on a wooden box covered with a Persian carpet and a pillow, embarrassed to be so singled out, while the captain, Abdul and five crew members crouch on the deck around me, chatting in Arabic which Abdul translates into English for me.

"The ship and all that is in it belongs to you," the captain tells me graciously.

"Thank you," I answer cheerfully. I try to return his hospitality as well as I can. "I would like you to visit me in my home in Toronto. I will give Abdul my address so you will know where I live." The vision of the Arabs in their long robes marching up Yonge Street delights me.

"We are grateful," the captain says solemnly.

From the conversation I learn that from two hundred and fifty to five hundred *dhows* sail to Africa each year, although the number is decreasing. The year before, twenty *dhows* were lost in storms or on shoals.

The crew of the Arab dhow I visited.

An Arab cook with an unwashed, cotton-print cummerbund hands the captain, Abdul, and me small cups without handles into which he pours thick, black coffee from a brass coffee pot. He also offers us sweetmeats—brown, putty-like masses from which we pull off bits to munch. The crew sit quietly, watching us chew.

The coffee is incredibly strong. I gulp it down finally, smiling determinedly as I do so. As I put my cup on the deck, I shake my head to show I don't want any more. I've had enough. The cook, smiling, refills my cup. I drink this cup too, trying to be polite, trying to look delighted. It's even stronger. I watch Abdul this time. As he hands his cup to the cook, I hand mine over too. To my horror, both cups are filled again. I drink down this cup too with increasing difficulty. It would surely be rude to leave it, especially when the crew has had nothing. Full of caffeine, I'm becoming desperate. What is the Arabic rule for politely refusing? I haven't the slightest idea. Abdul finally notices my distress.

"You've had enough coffee?" he asks.

"Yes, thank you," I reply firmly.

Abdul laughs. "You like it?"

"Yes indeed, but I've had enough."

"Then shake the cup as you return it. Like this." Abdul shakes his cup as he gives it to the cook. The cook doesn't refill it. I shake my cup the same way with equal success. Abdul and I exchange a look of triumph.

When the crew members have straightened their turbans and smoothed down their skirts, they line up in a row so I can take their picture. Then, shaking hands with the captain and thanking him, Abdul and I climb down the shaky ladder to the dock, Abdul going first and I clutching my skirt around my legs as best I can in case he glances up and sees my underwear. We exchange addresses, and when I send him a Christmas card almost a year later he replies, stating that he is "overwhelmed with joy due to the fact that you did not forget a friend in Zanzibar." He writes that on his yearly leave since my visit, he toured the Middle East and the Sudan, travelling overland from Cairo to Khartoum, later by riverboat to Juba, and as far as Uganda. The highlight for him was meeting members of the Dinka, Shilluk, and Nuer tribes in southern Sudan. Is such a trip a version for Arabs of the Grand Tour that so many Europeans and North Americans have made through Europe over the centuries?

Before we part Abdul introduces me to Bill, a European who manages a nearby clove factory. Bill gives me a tour of his plant, where oil is squeezed from cloves.

"Zanzibar produces 80 percent of the world's cloves," he tells me proudly.

"Do you like living In Zanzibar?" I ask.

"Very much. We don't overwork here. Zanzibar is a relaxed place."

"There aren't many Europeans here, though."

Bill laughs. "Most of them despise Zanzibar. It's a backwater, it's hot, and it's dull. But I love it."

I leave by plane the next day for Tanga, the nearest town to Zanzibar on the mainland, and the terminus of a train going west to Moshi, the town nearest Kilimanjaro. At the train station there are hundreds of people waiting, all Africans but me. They cluster near one end of the platform and I near the other, where the first-class cars with sleeping accommodation will stop for loading. The station master is upset to see me standing in the sun, so he lugs out a chair from his room. I point down the platform where, among heaps of luggage, dozens of exhausted-looking mothers stand clutching children in their arms.

"They need the chair far more than I do," I try to explain to the station master, but he doesn't understand. He becomes upset when I don't make use of the chair, so I finally sit down to be polite. I feel ridiculous resting in comfort while the other women stand. To take my mind off this dilemma I rustle up a pen and paper and begin to compose a letter home. Soon several men sidle up behind me to watch me writing; when I glance up at them they look sheepishly away, guilty to be caught staring. Is it because I know how to write? Or because they haven't seen English, as opposed to Arabic, written before?

The train finally leaves at dinnertime, myself ensconced in a compartment sleeping two—my roommate will be a student hairdresser boarding at Korogwe. I have drinks with an entomologist named Peter Walker, who works in Arusha. When I tell him my plan to visit there to see if I can arrange to study giraffe, he's excited.

"When you come to Arusha I'll introduce you to a Canadian, John Armstrong, who hails from the University of Western Ontario. It's not far from Toronto, I think," he says. "He's working with mosquitoes." What a small world!

When I wake up the next morning, my roommate and I both stare out the train window in awe at the distant, snow-capped peak of Mt. Kilimanjaro, towering three miles into the air. Can I really climb that high? Because that is what I plan to do. Can it really be snowy-cold up there, when it's so hot down here?

# 12

## Up Kilimanjaro

I leave the train at Moshi and share a taxi to the Marangu Hotel at the foot of Mt. Kilimanjaro. This hotel is one of two mentioned by John Gunther, a former classmate of my mother's at the University of Chicago, in his book *Inside Africa*. It advertised that it fitted out expeditions to climb Kilimanjaro, at over nineteen thousand feet the highest peak in Africa. Ever since I read about this the year before, I've wanted to take part in such an adventure. I wonder how many of us there will be on the ascent, and hope that some of them will be Canadians.

At the hotel I find, to my surprise, that the expedition clientele consists only of me, since the only other prospective climber has backed out; this apparently isn't as popular a trek as I had imagined. However, for only $55 Canadian, which includes food, warm clothing, a sleeping roll, and five men to convey the gear and me up the mountain, this seems a good buy. The first three and a half days of the trip will be spent struggling thirty-five miles forward and nearly three miles up; the last day and a half will cover the return from the summit to the hotel.

Real mountaineers dismiss the climb up Mt Kilimanjaro as not difficult, offering only the challenge of distance and altitude. To laypeople like myself, however, seventy miles is a long way to walk and nineteen thousand feet an altitude with too little oxygen. Indeed, the hotel manager tells me that fewer than half of those who begin the climb ever reach Gillman's Point on the rim of the crater at the top.

The evening of my arrival at the hotel, the manager helps me choose clothing for the ascent. She is not impressed with the wool blazer I've lugged through Africa with this trip in mind, nor with my anti-sun precautions.

Instead, she goes to the storeroom and comes back with various substitutes and additions for me to take: sun-goggles to replace my sunglasses, a second jar of face cream, a bigger hat, two scarves, mittens, two sweaters, a balaclava, and a pair of corduroy jodhpurs to wear under my jeans.

"It's very cold towards the top," she says, "so you won't need pyjamas or a change of clothes. You'll just keep adding more layers the higher you go." I realize the extra wisdom of this remark later when I find that men and women all sleep together in the same small huts. She gives me a first-aid kit, explaining that aspirins are particularly important as many people have severe headaches all the way up the mountain. Others can't eat, or vomit constantly.

When my belongings have been completely revamped, the manager leaves me with the journal in which previous climbers have written about their trips. The pages are crowded with tales of mountain sickness, headaches, broken legs, and pneumonia. All the notes warn the beginner to go slowly. One man warns me not to keep on if my heart is beating too loudly for too long. Another writes that it is impossible to really damage one's heart by such exertions, because one will fall down first from exhaustion. I learn that after each day's climb of about eleven miles, there are permanent huts in which to sleep. The final climb to the summit from the third and highest hut begins at 2:30 in the morning, with part of the thirty-five mile descent taking place later that day. Many of the climbers end their message with the words "Never again."

In the lounge, before going to bed, I ask the manager about the Chagga tribe who live in the area.

"They're wonderful people," she says, "probably the most progressive in Africa. This part of Tanganyika in general is very dry, but by irrigating their crops with water running off the mountain they're able to grow maize, bananas, and even coffee bushes."

After breakfast I meet my guide, Fernandes, a WaChagga who speaks even less English than I speak Swahili. I try to appear strong and durable as he looks me up and down without smiling. He's wearing baggy brown pants, with a torn, white, sleeveless sweater over a striped, short-sleeved shirt. His hair is so short as to be barely visible.

"The porters have already gone on," the manager explains to me. "They've got all the food and bedding and your clothes. You'll meet at Bismark Hut this afternoon." Actually, I never do meet them officially, as Fernandes never introduces them and struggling to chat with their client seems the farthest thing from their minds.

Fernandes and I set off about ten o'clock, each carrying a knapsack and an alpine stick. We walk for a mile up the paved road, Fernandes chatting to the Africans we pass. They all answer him, then stare at me and laugh. I say *jambo* or "hello" to those who look as if they'll listen; a few answer *jambo,* then laugh again. As we turn right off the road onto a dirt track there is a modern, one-storey schoolhouse, where small African boys and girls dressed in grey-brown shorts or jumpers are holding hands in the yard, marching backwards and forwards, and singing in high, sweet voices while directed by two African teachers.

Now our path leads steeply upward between the *shambas,* or small holdings where corn plants and banana trees grow, both irrigated by water running down the mountain in a nearby stream. Among the *shambas* are small dwellings, most of them round and shaped like beehives covered with overlapping layers of thatching grass, but a few new ones are constructed of cement blocks.

Occasionally we pass a tiny *duka* or store surrounded by lounging Africans. I'm invariably recalled, reluctantly, to my own culture by a red Coca Cola sign on a nearby tree. We meet a number of women on their way down to market with huge straw baskets full of corn cobs on their heads. Others balance bundles of firewood or thatching grass. They all bound past in their bare feet, trying not to stare at me as they exchange comments with Fernandes. The women wear brightly coloured cotton dresses with large, coloured kerchiefs like table cloths knotted around their necks and flowing down below their waists. Sometimes Fernandes stops to greet a friend; they shake hands and converse for a moment while I stand by awkwardly, wondering what they're talking about.

Soon we meet the four porters resting under a bush, dressed in ragged shirts and shorts, all but one with bare feet. After Fernandes and I relax for awhile with them, the men chatting among themselves, we all start to climb again. Two porters carry green wooden food boxes, one with a bundle of clothes on top of his box. The third porter carries my clothes wrapped in my bulky bedroll (comprising a pillow, mattress, and four blankets) and the fourth a huge gunny sack bulging with unknown contents. Each porter puts his load onto a doughnut-shaped pad of fresh banana leaves placed on his head. Two of the men also carry lamps, and two have whisky bottles hanging from their shoulders by a string, each full of kerosene for the cooking fires. Fernandes, who up until now has carried sandwiches and a thermos full of tea in his knapsack, perches a dunnage bag and bedding on his head and picks up a Primus stove in a basket to carry over his arm. I continue to carry my camera, a notebook, and a pen in my knapsack. Fernandes walks with me each day while the porters, who keep their own time, are usually far ahead of us.

Leaving the prosperous green *shambas* behind us, the track becomes a path leading into the dark rainforest which originally covered the entire base of the mountain. At the edge of the forest a few Africans are cutting firewood, but soon we're above all sounds of humanity. The vegetation around us is dense with a profusion of creepers and epiphytes, so it's impossible to see far on either side of the path. As it begins to rain lightly, increasing the gloom in our forest tunnel, we meet a European couple and six porters coming down the trail. The Europeans and I greet each other joyfully, the couple stopping long enough to tell me how awful their trip has been. The man hadn't been able to climb farther than the second hut, while the woman had encountered a blizzard before reaching the final hut and had been too sick to carry on. They wave goodbye as they depart, euphoric to be nearly down. Fernandes and I continue to trudge upward.

At first Fernandes checks my enthusiasm by saying *pole pole* (to rhyme with "holy")—go slow—but soon his advice is unnecessary. When the path is very steep, I'm convinced I'll never get to the summit; on the more level parts I stop thinking up excuses as to why I've failed and become determined that I will never give up. Three European men in army jackets and without a guide or porters swing by us going downhill, the first growling hello and the others not deigning to speak. Fernandes and I exchange smirks—they probably didn't make the summit or they wouldn't be so glum. Soon we meet another couple who were caught in a hailstorm and hadn't even reached the last hut; they don't stop to talk either.

I'm delighted to halt for a lunch of sandwiches and tea, which Fernandes and I share beside a small stream in the dim forest light. We chat sporadically, using one of the few words we have in common—*mzuri,* Swahili for good or well. Sitting down for lunch is *mzuri,* the rainforest *mzuri,* the tea *mzuri mzuri,* and the time to go again *mzuri.* We accompany each *mzuri* with a suitable nod or gesture.

After lunch we struggle on up the path. Gradually, as the forest becomes less dense, we traverse several open glades with grass, bracken, and blackberry-type vegetation. Sometimes brightly coloured flowers are scattered in open meadows. I'm beginning to think we'll never stop climbing when we arrive at Bismark Hut, named by the Germans who colonized Tanganyika before the First World War, nine thousand feet up and about twelve miles from the hotel; it's a handsome one-storey building made of stone, with shuttered windows and a steep metal roof. Beside it runs a small stream from which the porters collect water for drinking and cooking. They have already made themselves at home in their own hut, and bring me a welcome glass of hot milk and a plate of six cookies.

It's very cold, especially now that I've stopped walking, so I put on a scarf and mittens and crawl into my bedroll on a wooden bunk to read *Crime and Punishment* until suppertime. My bedding is an odd mixture of luxury and austerity; although it's wonderful to have a mattress and four blankets, I'd rather have forgone the embroidery on my pillowslip for a second sheet to keep the blankets from scratching.

Before long a couple from Nairobi arrives who have walked the thirty-two miles from the summit that day. They are too exhausted to do anything but talk.

"The final ascent this morning was ghastly," says Clive. "I was sick to my stomach, but kept on anyway."

"It was freezing," chimes in Diana. "I've never been so cold in my life."

"But the crater was marvellous, well worth the whole trip." They are very pleased with themselves.

Clive lends me his strong boots and Diane makes me a Swahili vocabulary of useful words such as "cold," "hungry," and "latrine." Then, in a festive mood, we eat dinner together by lamplight. The bleakness of the three-room hut is not reflected in the food any more than in our spirits. I'm given soup, a plate of meat, spaghetti, potatoes, cauliflower, carrots, bread and butter, a dessert of bananas and oranges, and a white linen napkin in an embroidered holder. This feast is served on a white tablecloth spread out by one of the porters.

The next morning, Fernandes and I say goodbye to our friends and set out at 8:30 after a sumptuous breakfast of half a grapefruit, two bananas, porridge, liver, bacon, two eggs, bread, jam, and coffee. For the first mile we climb up through a dense, wet forest. Then suddenly we're out on the meadows, surrounded by fields of tall grasses and flowers clouded in mist. Above the mist the air is so clear that we can see our destination, snow-capped Kibo Peak, many miles away. Fernandes, pointing to a heap of old elephant dung, says *tembo* proudly, as if he's guiding a tour. The path is soon a succession of boulders, roots, and puddles, twisting relentlessly upward. I try to justify my halting progress to Fernandes by muttering *pole pole* now and then. Around us the grasses give way to giant heather plants, huge lobelias, and scattered aloes.

Early in the afternoon we arrive at Peter's Hut at twelve thousand feet, a small shack with gray tin walls and a rusty tin roof, which I find I'm to share with a party on its way down the mountain. The group is composed of an English couple, an agile young Scotsman who has not only climbed to the crater rim but part way around the rim, and a man from Edmonton, Alberta. A man from Canada!

Anne at 12,000 feet altitude on Mt Kilimanjaro with the summit beyond.

"How was it?" I ask him with excitement.

"Awful," he says with a dead voice. "On the last stretch, when we weren't that far from the top, I suddenly knew that the pleasure of the summit couldn't possibly be worth the agony of getting there. Anyway, I don't want to talk about it."

"It was just too much," says the woman, who's wearing a skirt. "We were all just too exhausted to go on. But you're younger. I'm sure you can make it. Don't give up like we did!"

Again we clients stay in the main hut while the guides and porters share two others. We spend the afternoon wrapped up in our blankets lying on our bunks, the extreme silence in the hut broken occasionally by grunts from the Scotsman as he shifts his position on the wooden planks above me. I watch a brown mouse with two white stripes down its back foraging cautiously for crumbs under the table. Outside the hut are some black carrion crows, a few small brown birds with long bills, and a number of brown striped lizards, the last animals I see on the ascent. Toward evening the mist clears to reveal the snowy summits of the two peaks, Kibo and Mwenzi, sharply outlined in the distance against the blue sky. I decide I want to reach the top of the higher peak, Kibo, more than anything else in the world.

After another elaborate dinner we sit bundled around the wooden table, chatting cursorily, aware of the more animated chatter and laughter of the Africans a short distance away (one has complained of a headache so I have given him two aspirins, pleased to be reminded of the gentry's noblesse oblige.) The oil lamp throws distorted shadows into the corners of our hut and our words seem lost and forlorn. I retire to bed again at eight, but lie awake for hours, listening for sounds to break the absolute stillness of the night. My fellow climbers are awake too, because at intervals one sits up and rearranges the bedding in an attempt to combat the cold.

On the third day's march, I'm aware of the effect of altitude for the first time. In the morning I set off briskly up the hill behind Peter's Hut, but before I've gone thirty yards, I feel such a sharp pain in my chest that I can hardly breathe: a terrifying sensation. Will I have to give up? When I've rested a minute, though, I can still climb if I move at a snail's pace. If I creep any faster, my heart twinges every time I breathe. I soon learn to take a huge breath in during one tiny step, then breathe out for two even tinier ones. At this pace it takes seven hours to climb the ten miles to the last hut at the base of Kibo Peak, so I have plenty of time to look about and enjoy the scenery.

For the first five miles we creep left around the base of Mwenzi, a superb peak of rocky crags towering above us, etched in snow. It's here that an airplane crashed two years before, a few remains of metal still reflecting the sunlight. The vegetation becomes shorter and sparser as we progress until only scattered bunches of grass, small flat clusters of tiny flowers, and lichens remain. When I have to urinate, there's nowhere to hide. Fernandes does so after taking a few steps aside and turning his back to me; I wave my arms to indicate to him I'm leaving the track for a short distance, hoping he won't look my way as I crouch down.

Before noon we meet a European woman and two men with their porters coming down the path. The couple refuses to speak to me so I'm pleased when Fernandes gestures that their summit attempt has failed. Of the sixteen people who pass me going down the mountain, only three have reached Gillman's Point.

On the saddle stretching between Kibo and Mwenzi Peaks there is no vegetation at all, only brown gravel extending to the horizon in every direction with occasional large, brown boulders scattered about—a magnificent if desolate sight. Fernandes and I eat our sandwich lunch with our backs propped up against one of these large rocks, where we can watch the porters ahead of us growing smaller as they cross the brown desert to Kibo Hut. Their energy level seems little impaired by the altitude; although they have now added bundles of firewood and large cans of water to their loads, they march along steadily without pausing to rest.

Fernandes and I compare wristwatches, writing down on a slip of paper the number of shillings each has cost. Fernandes's is the more expensive, which is ironic and pleases Fernandes, and which seems democratic and so pleases me. I spend some time trying to explain in sign language that mine is ten years old, but I don't succeed. Fernandes finally nods doubtfully at me to be agreeable.

When we have finished lunch I stand up to put on my knapsack, but this slight exertion makes my heart beat so wildly that I sit down and rest again before actually starting to walk. I never imagined that anyone could travel as slowly as we do. Although Fernandes moves like a tortoise, he is always far ahead of me. I take about one step a second, with my front foot never more than a foot's length in front of my back one. Several times I feel like crying with discouragement at my lack of progress, but sometimes I burst out laughing in the best of spirits and chuckle to myself as I creep along. Is this swing of emotion an effect of the altitude? The only living sight now in this brown world is Fernandes, far ahead of me, the only sound that of the cold wind blowing incessantly across the lava plain. This last part of the trek is over scree, or small stones, sloping gently upward. Ahead of us towers Kibo Peak, a series of sweeping grey valleys merging upward into a tremendous flat plateau that is the crater rim.

Kibo Hut, at sixteen thousand feet, is even smaller and more primitive than the earlier huts, reflecting the reality that the higher one goes, the fewer clients remain to be catered for. It is only ten by fifteen feet square, surely one of the most isolated habitations in the world. There is no life visible anywhere and the wind whistles round the hut and through the cracks in the walls. The evening is fresh and clear, and colder than ever. Once, forgetting how high I am, I stroll over to an unusual stone not far from the hut. Immediately, my chest is so crippled with pain at this sudden activity that I have to bend over and gasp for air.

How lucky I am to have the mountain almost to myself! If it were forty years later I would be one of twenty thousand people clambering each year up the mountain along a variety of routes. In 2005, on any one evening, there could be dozens of tents outside Kibo Hut making up a large tent city, the air no longer fresh but redolent of overwhelmed outhouses.

I eat my dinner on my bunk because the hut is too small to accommodate chairs. Before settling down for the night, I read entries of climbers written in an exercise book kept in the hut. They are anything but reassuring:

- Misery of miseries, Hell of hells
- A ruin, a hell, a misery but worth trying
- Never again
- This is really the worst day that I have ever had
- Have never suffered so much discomfort and cold in my life
- To serve, to strive, and not to yield.
- NEVER AGAIN.

Trying to sleep at sixteen thousand feet is as bad as trying to walk there.
I find to my consternation that I can only breathe if I lie flat on my back in
order to give my lungs as much space as possible. It's bitterly cold, my muscles ache, and I feel as if I weigh three hundred pounds. I shift my weight onto
one hip until it's sore, then shift it onto the other. I can hear two imaginary
voices, one purporting to be Kibo Hut and the other Peter's Hut. Kibo Hut
urges me at great length to lie on one hip; then Peter's Hut puts forward reasons why I should lie on the other. I'm glad when an imaginary American couple enters the room to converse between themselves and with me, because then
the huts become less vocal. I open my eyes several times to see if the couple
really is imaginary. Because I can't sleep, I'm also glad that we will be getting
up very early—essential so that the ground will remain frozen as we climb to
the summit.

A porter arrives at one a.m. to rouse me; the others have descended to
Peter's Hut for the night. He's dressed in a khaki coat with a high, pointed
hood that makes him look like a Tibetan monk. I lie and watch him as he
moves quietly about in the shadows. As he starts the fire by pouring kerosene
into the stove, I wonder vaguely if we'll be blown up. It seems bizarre to be
in such a cold, forsaken spot in the middle of the night with only a few African
men I can't talk to for many miles.

After the porter makes me a cup of tea (I refuse the offer of porridge—
ugh), he and Fernandes sit down in the dim light to wait for me to get ready.
I want to rearrange my layers of clothing, so I comb my hair and putter about
while the two of them watch. It's several minutes before I realize that they have
no intention of leaving the relatively warm hut. By waving them to the door,
opening it for them to go out, and closing it after them, I manage to be alone.
The outhouse is about twenty yards downhill from the hut, so I have to lean
over and gasp when returning from it, making me doleful about the coming
enterprise.

At 2:30 a.m., Fernandes and I set out for the top in the cold night air, the landscape flooded with moonlight. Behind us, Mwenzi Peak stands out magnificently against the dark sky. Fernandes is bundled up with a scarf, balaclava, and long coat over his war-surplus windbreaker. I have on three pairs of socks, two pairs of trousers, three sweaters, two pullovers, two scarves, mittens, a balaclava, and sun goggles, with my trench coat on top of it all. We both carry knapsacks and alpine sticks tipped with metal to dig into the gravel scree.

The first two miles are over a gradual slope upward, so we climb steadily at a snail's pace. After that the scree becomes steeper. My feet slip on the gravel so that my creeping method involves nearly as much sliding back as it does moving forward. I try taking little dashes and then stopping to gasp before attempting another spurt. My heart no longer bothers me, but I have trouble breathing. My legs become tired quickly; sometimes, losing control of them, I lurch off the path. I'm glad I don't know at the time that we are approaching the altitude, 17,700 feet, at which the Everest *base camps* are situated. Imagine *beginning* a climb of over ten thousand feet at an altitude where the air has so little oxygen!

After struggling on for what seems an eternity, we stop at 4:30 to rest in a small cave about halfway to the summit, although it's too dark to see how hopelessly far we still have to go. The path is now very steep, the scree more slippery than ever. After each small dash I sit down to recover. The sky is lightening behind Mwenzi Peak, so I pretend that I'm resting to watch the dawn. I doubt if I fool Fernandes. Progress is so incredibly slow and agonizing that, finally, I decide I can't go on. I would have burst into tears of despair, but I feel much worse than that. I look hopelessly at Fernandes above me and shake my head. I try to think what "I can't go on" is in Swahili. Fernandes only grins companionably down at me, jabbers something, and continues climbing. I'm too beaten to do anything but follow.

The climb becomes a kind of hell; I'm cold, tired, breathless, only able to take a few steps at a time, and the earth ridge above etched against the blue sky never seems any closer, hour after hour. The air is so clear at this height that the top looks fifty yards away, but after struggling for several more hours it appears no nearer. When I ask Fernandes how much farther it is, he holds up three fingers indicating three miles? or three hours? I'm sure he has to be mistaken; no one could possibly last for three hours more of such torture, let alone three more miles.

If I go slowly I can cover three yards at once. After each three-yard stretch, I rest for two minutes until my heart stops pounding and I can breathe instead of gasp. While I rest, I choose a stone to mark the end of my next spurt. Sev-

Porters resting.

eral times, when my choices are too ambitious, I can scarcely even gasp after the lap.

Eventually we stop for tea. It's now daylight, but still too cold for us to sit down for long. My pace decreases to one yard or so a minute when we start again, so that even the patience of Fernandes gives out. He takes hold of one end of my belt and half pulls me up the last fifty yards, stopping every eight yards or so for me to catch my breath. I push myself from behind with my alpine stick.

Suddenly, with one final spurt, I'm on the crater rim with only air around me. Fernandes and I grin at each other with glee. What luxury to be able to stop! Ahead lies a mighty hole, behind, the endless saddle. We did it! Yet the crater, our destination, is an anti-climax after the last three days. No sign of the leopard made famous in Hemingway's story "The Snows of Kiliman-jaro," who is thought to have come up hunting for eland. Although the crater's diameter is over a mile across, the depth seems only to be a few hundred feet. It encloses several cliffs of snow, but much of its floor is bare. The view out over Tanganyika is hazy, but the view over the saddle is a wonderful stretch of shaded browns and rusts, with tiny, white patches of snow. Kibo Hut looks unbelievably tiny. Nearly fifty years later, Kilimanjaro's snow cap will be gone, a casualty of global warming and a portent of the environmental crisis that awaits us.

After signing the book kept at Gillman's Point at nine o'clock, "I Made It," we start the rush down the mountain, digging our heels into the scree at

Fernandes and Anne, successful conquerors of Mt Kilimanjaro.
Anne wears the victory wreath made for her by a porter.

each leap. It's magnificent to be able to move and breath normally, although my legs are shaking when we reach Kibo Hut. It has taken us forty-five minutes to come down the path we climbed up so painfully for six and a half hours. The porter grins at me for the first time and gives me a whole tin of apricots to eat, one of the most welcome repasts I've ever had.

The walk back across the saddle is wonderfully invigorating. As we stride along I feel that now I can do anything in the world. When we come to the slightest incline of the trail, though, my heart still protests immediately; it's not until we are well below ten thousand feet that I can ignore the slope of the path. We reach Peter's Hut at two o'clock, where the porters are camped, anxious to get home, so we decide to carry on to Bismark Hut. I'm still too elated and adrenalin-infused to feel tired, so ten more miles seems like nothing after what we've been through. The afternoon march is difficult, however, because rains have left puddles and made the rocks on the path slippery. In my hurry to keep up with the porters, I bang my ankle on a boulder and fall down in the mud.

At Bismark Hut are two New Zealanders, Peter and James, on their way up the mountain. They regard me apprehensively as I limp in covered with mud, my ankle swollen, and my feet blistered. When I try to clean my hands, I realize I have no soap and haven't washed for four days. The two men are wearing cashmere sweaters, white knee socks, and immaculate shorts and shirts. They urge me not to change for supper—easily done—and one of them apologizes because his sweater has little bumps at the shoulders where he has pegged it up to dry after washing it. They know no Swahili at all, so I make out a vocabulary for them of the ten words I've mastered. They agree to write to me after they've reached the crater so we can compare experiences, but I never hear from them.

On the following day, before we set off, the porters present me with a wreath of alpine-type flowers to wear on my hat for having reached the top, a souvenir I still treasure fifty years later. Hot and sweaty, we arrive at the hotel at lunch time, where I shake hands with Fernandes and the porters, and say goodbye.

"Can I buy them each a drink?" I ask the manager, who has come out to congratulate me on the success of the trip. I'm not sure of proper protocol, but want to thank the men for their help.

"Just give them money for a tip," she suggests, which I do.

"Thank you for the adventure of a lifetime!" I tell them exuberantly. They don't understand, but they smile anyway as they pocket their money and turn to go home.

# 13

## *To Study East African Giraffe?*

The afternoon of my return from the mountain, I borrow the hotel washing machine to launder the sweat and mud out of my clothes, which I then spread around my hotel room to dry. This may sound like a simple chore, but I'm so stiff that, although I can stand, I can bend only with the utmost pain. If I want to sit, I fall backwards toward a chair, collapsing all my muscles on the way down. To get up, I pull myself to my feet by grabbing a chair, table, or other nearby piece of furniture. My big toe nails are sore, too, and beginning to blacken from the constant battering they've suffered hitting the front of my boots on the descent down Kilimanjaro. I spend the evening lying immobile on my bed, reading *Crime and Punishment.*

The next day I write letters before lunch to Ian and my mother, recounting my adventures on the mountain—my mother tells me later that she and the whole Women's Union where she lived were in a state of suspense, knowing I had started out "alone" on this trek but not knowing if I would return safely. I then pack my suitcase and knapsack and totter to the main road to wait for the bus to Arusha, thirty miles away. From there I plan to organize my study of giraffe in East Africa now that I've achieved my dreams of visiting Zanzibar and of climbing Mount Kilimanjaro. Although Arusha is the headquarters of the Serengeti National Park to the northwest, this immense, newly established park is scarcely organized. In the month before my arrival, only twenty-two permits were issued for visits; by the next century, over ninety thousand tourists will enter the park each year.

The bus is to pass at 3:30, but at 4:30 I'm still waiting beside the road, standing in the sun. I'm not sure what to do—is the bus just late? Or has it been cancelled, which means I'll have to spend another night at the hotel?

Then a small car drives past me and stalls, or pretends to stall, at the side of the road. In it are two policemen training at Moshi: Ron from Australia, who has a headache from too much Canadian rum drunk the night before, and a South African, who cares nothing about politics (always my first question to South Africans). They know Paddy, Julie's boyfriend, and John Cairns from Dar es Salaam.

The men are going to tea at the Kibo Hotel so I go with them in their car, which turns out not to be broken down after all. It's two miles further up the mountain than my hotel, surrounded by rainforest, and with caged monkeys beside the road. After tea, the two men obtain permission from police head-quarters to drive me to Arusha, which they do. This gives them eighty extra miles of driving, but that doesn't bother them in the least. Perhaps there isn't much crime to fight in Moshi?

I check into the New Arusha Hotel, a handsome one-storey building with curtained windows on whose lawn an impressive sign reads "This spot is exactly halfway between the Cape and Cairo and the exact centre of Kenya, Uganda, and Tanganyika." Wow! The next morning I set out on foot to see the town, overlooked by Mount Meru towering behind it. The first person I meet by accident is Peter Walker, my train acquaintance, who takes me imme-diately to meet my compatriot John Armstrong, the Canadian working on mosquito research. John is a classmate of three biologists whom I know because they came to the University of Toronto to do graduate work. I gather there are few single white women in Arusha, because John is keen for us to meet in the afternoon for drinks at my hotel.

My next encounter on the street is with Don Longlons, whom I had met on the *Kenya Castle* sailing to Dar es Salaam. What a small world! He drives me to see the old German Fort, its immensely thick walls newly whitewashed, then to his house for an orange squash, a popular drink made by adding cold water to a concentrated orange liquid.

In the late afternoon I wait for John in the hotel bar, where a man points out Ernest Hemingway's son entertaining several women with bleached blond hair, all drinking and laughing loudly. A safari party of three older couples, decked out in brand new khaki shirts and trousers, with topee hats and mos-quito boots, sits near me chatting with their two White Hunters. I can hear occasional, titillating snatches of tales involving lion chases and close encoun-ters with snakes.

When John arrives, we reminisce about Canada for a while before I tell him about my plan, if possible, to do research on East African giraffe from a base in Arusha. "I have a letter of introduction to G.H. Swynnerton, the Chief Game Warden of Tanganyika," I boast.

"That's great!" he exclaims. "There are lots of giraffe in the area; safari companies actually use Arusha as their base of operations. It's a small town, so easy to get around. Less dangerous than Nairobi, too."

"Could you help me get organized?"

"I'd love to. Your best bet is the Game Department at Tengeru, where Swynnerton works," he says. "Their job is to keep down poachers and foster tourism, but they must know a lot about giraffe, too."

"Where's Tengeru?" I ask.

"Just outside Arusha. I'd drive you, but I have to work."

"Is there a bus?"

"No, but you can take a taxi," he replies. "They don't cost much, but ask about the rate before you get in because they don't have meters."

Before he leaves, he gives me a kiss. Except for Ian I've never been kissed on a first date before, but I don't mind because I'm excited about the prospect of settling into more giraffe work.

The next morning I phone Tengeru to explain that I'd like to research the behaviour of giraffe in the area and to make an appointment to discuss this. "Mrs. Nicholson from the Labour Department has written to Mr. Swynnerton and to Mr. Thomas to tell them about me," I say, determined to sound important.

"Mr. Swynnerton and Mr. Thomas aren't in at the moment," the receptionist says. "Would you like to meet with Mr. Maintz?"

"I guess so," I say; I have no idea who this new man is. "I'll take a taxi right over."

The office of the Game Department is a small, one-floor building set along the main road where the African taxi driver decides to wait for me, even though I'm not sure how long I'll be.

"I'm afraid Mr. Maintz must have gone out," the receptionist says apologetically when I ask for him. "I told him you were coming, but now I can't find him."

"I can wait," I say, annoyed, thinking of the taxi cost.

"He probably won't be back today," she says.

"I'll wait for Mr. Swynnerton or Mr. Thomas then."

"Actually, they've gone to Nairobi and won't be back for days."

So much for the Game Department. I take the taxi back to my hotel and pay the one pound fare. Damn. From the taxi I've seen a sign proclaiming "Tanganyika Safari," so I decide to go there to see what prospects it might have for giraffe research.

None, it turns out, although the fifty-year-old White Game Hunter boss, Russell Douglass, gives me tea and regales me with tales about animals he's encountered in the wild—mostly lions and elephants. Near the hotel I drop in to a gift shop to buy a Masai anklet where I meet another White Hunter, young this time, named John Seed. He introduces himself to tell me not about giraffe, but about his hangover and his many girlfriends: shades of Ernest Hemingway. Another White Hunter is in a back room counting crockery in preparation for his next safari with rich Americans, but he cares nothing about giraffe either.

In the late afternoon John Armstrong and I visit the pleasant home of Peter Walker and his wife, where we look at photographs of Africa and a few of giraffe. Then John and I go to dinner and to dance at the Beehive Bar where John Seed, quite drunk, is chatting up two schoolteachers and a British woman, Mary Bradley, who is working her way around Africa. John and I kiss more seriously as we say goodnight.

In the morning I accompany John on the seven-mile trip to the Colonial Pesticide Station, where he works in a spruce new building along with physicists, chemists, and other biologists. His job involves cruising around the country in his company car collecting mosquitoes from various sites, then breeding colonies to see how various pesticide sprays affect them. His main focus is on combatting malaria, which is spread by mosquitoes.

In the afternoon we drive to Tengeru—still no men in charge unless they're hiding—to obtain a permit to visit the Ngurdoto Crater, about twenty miles from Arusha. The route is along a dusty trail, first through dry grassland and bushes with some zebra in the distance, and then up through dense forest to the crater lip, which overlooks a miniature world spread over several miles of crater floor, far below. With field glasses we see nine tiny giraffe, one rhino, about forty elephants, and two hundred buffalo. There's no way to study giraffe here, though, when I can't get near to or even see many of them among the trees below.

On the way home we stop at the White Hunters' Game Farm, which is just getting organized. It has a variety of baby animals which it plans to sell to zoos around the world, but only one giraffe: a baby who licks my hand, then tries to bite my fingers. No research future here, either, although John says, hopefully, "I'm sure they'd let you come here every day to study this little fellow."

The next day we drive to the Keunzler Game Farm near the Pesticide Station. This is a private game farm started up by Mr. Keunzler on his own property, where there is a huge shaded house, a swimming pool, and a beautifully kept lawn where a dog and antelope rest side by side. The Game Farm inhabitants include a large and a small giraffe, baby elephants, rhino, wildebeest, and lots of ostrich running loose. A mouse runs out of a bale of hay near us which an African man immediately chases, cutting it neatly in two with his *panga,* or machete. At least there are two giraffe here rather than only one, but, again, they live in zoo conditions. There's no way to study wild giraffe here, either.

In the afternoon—John seems to have lots of spare time despite his work—we drive up the west side of Mount Meru for the view. We stand romantically side by side, holding hands, gazing over the magnificent vista of pasturelands and rolling volcanic hills disappearing into the distance. After several minutes, becoming bored, I peek at John without obviously moving my head; he is gazing steadfastly ahead, shifting his head in a slow arc to take in every bit of scenery, so I keep looking ahead too. Soon I stop focusing on the panorama and wonder who will break free first in our apparent viewing deadlock. Can he really like hills this much? Does he see something I don't? Eventually, finding the increasing silence unbearable, I break the silence.

"Gorgeous," I say, squeezing his hand.

He starts, swivelling his attention toward me. "Yes," he says, smiling. He's won the aesthetics non-contest.

We have supper at his house, prepared by his two male servants; a third African "boy" who works for John's roommate is also busy around the bungalow. Having his own servants while in his twenties is quite a step up for a small-city Canadian boy. I would have been chary of going to a man's house by myself in the evening, but with so many other people about there's no danger to my virtue.

The next day we're up at sunrise to drive southwest to the Rift Valley, initially travelling over vast expanses of pastureland, sometimes with no trees at all visible as far as we can see. This region is degraded by large herds of cattle, but later we also see giraffe, eland, and Thomson's gazelle. We cook breakfast over a Primus stove, the food carefully packed in a wicker basket by John's "boy." A tall, gaunt, Masai man stalks past us, wearing only a rust-coloured blanket along with large hoop earrings and a heavy bead necklace.

At the top of the Rift Valley we stop to admire the vista below of Lake Manyara, parts of it solid pink from the hundreds of thousands of flamingo gathered there, and of the plains beyond. With field glasses we can make out fifteen giraffe, a rhino, and wildebeest but here, too, John admits there are no facilities to study the giraffe.

Two Jeeps owned by the Carr and Downey safari firm from Nairobi out on a one-month expedition, each carrying an elderly American couple, two African men, and a White Hunter, pull up to join us at the lookout. The wives study the view below with field glasses from the Jeeps while their husbands get out to move around.

John Armstrong on the rift wall overlooking Lake Manyara.

"Don't go too near the edge," one wife calls out several times, as if her husband is a child.

"What a great view," says her husband. "Maybe I could pot a buffalo from here?"

When the Jeeps drive off, two four-ton trucks fall in behind them, each loaded to the top with safari equipment and a number of African "boys." "The clients live like kings with Carr and Downey," John says, rather enviously I think. "They have gourmet meals on tables with fine dishes and cutlery, and they sleep in beds rather than cots."

"It must be nice," I say.

"They could set you up in style to study giraffe," he laughs.

"Yeah, right," I say. "All my money would probably buy me a week at most."

Arusha seems hopeless as a centre from which to study giraffe but the next day, grasping at a straw, I again phone the Game Department to see if Mr. Swynnerton or Mr. Thomas have returned from Nairobi. (I have no faith at all in Mr. Maintz—maybe he doesn't even exist?)

"They got back from Nairobi yesterday," the receptionist says cheerfully. "You can see them this morning if you like."

I grab another taxi outside the hotel—surely this time I'll be lucky—and do indeed meet Mr. Swynnerton in his office.

"I'll introduce you to Keith Thomas," he says when I ask about possible giraffe research based at Arusha after we've exchange a few pleasantries focusing on Mrs. Nicholson, Kilimanjaro, and the dry weather.

I imagine that Thomas must be the giraffe expert, given Mr. Swynnerton's behaviour. However, he turns out to know nothing about giraffe either, although I do find that he's a friend of a friend of John Cairns. He's embarrassed by his lack of knowledge about giraffe, the reason I've come to the department. He shows me several pictures of male giraffe, as if I don't know what they look like, and mentions that giraffe make no vocal sound, which I know is untrue.

"Are there any reports about what giraffe eat, or how often they drink water?" I ask. "Do you have any data about their distribution in East Africa? Or maps of the ranges of the various races?" I can see him pondering how to proceed when he can answer none of my questions.

"Giraffe often browse near sable," he says finally, and as he does so he rushes into the next room to unearth a picture of a sable antelope. We spend the next half hour looking at photographs of African mammals. Only after he has also shown me through a small museum featuring heads and skins of game animals does he feel that he has done enough for me, a visiting scientist. I leave the department thoroughly disillusioned.

Once back at the hotel I have lunch and retire to my room, disheartened. There seems no way at all to study giraffe from Arusha. John is off collecting mosquitoes so I don't even have anyone to talk to about my disappointment. I'm suffering from menstrual cramps so I crawl into bed for a nap, and have just dozed off when there's a knock at the door. It's John, back from work.

"Why are you in bed in the middle of the day?" he demands, laughing.

I tell him about the game department fiasco.

"Best to be up and doing, not hiding away," he chides.

"I've also got the curse," I concede in defence of my laziness, something I've never admitted to another person in my whole life, even my best girl-friends.

"Great!" John exclaims. "We can have sex right now 'cause there's no danger of you getting pregnant!"

I look to see if he is joking, but it seems that he is not, although he sniggers self-consciously when I stare at him. No way. Instead I have a prim tea with Mary Bradley and dinner with Frank Addison and his wife, friends of John Cairns to whom he has given me a letter of introduction. At nine p.m., John calls for me so we can go to see the new movie, *Annie Get Your Gun*. We kiss more seriously than ever as we say goodnight, as Doris Day trills "Que Sera, Sera" in the background; I'm leaving the next morning for Nairobi to see if I'll have better luck there in finding a way to study giraffe. In the 1950s there is sporadic debate in the media about the ethics of young couples kissing and caressing with abandon but never going "all the way." Some doctors said that such restraint was bad for the physical and mental health of both the boys and the girls. It was undoubtedly frustrating for both sexes. However, it did give young women a feeling of responsibility over their own bodies which many of them lost in the 1960s when, with the widespread use of birth control, sexual intercourse among the young became almost a cultural imperative.

After picking up sandwiches from the hotel kitchen the next morning, I take a taxi to the bus station. I intend to ride third class to save money and so I can hob-nob with real Africans, but the ticket master is reluctant to sell me a third-class ticket.

"You'll be happier in the second-class section," he insists, "or in first class, which has a bus to itself."

I decide he's right when I glance into the waiting bus and see the mass of humanity crowded into third class, sectioned off at the back of the bus. Even the second class is jammed with fifteen people, four of us "Europeans." I squeeze into a seat beside a white artist who lives with her sister on Mt. Meru; behind us on a three-seat bench crowd two Indian women with their four children, each of whom spends the trip either crying or resting between bouts of crying.

I leave Arusha with regret. I'll miss John and I'll miss its country setting and clean open streets. I'm glad I can't see into its chequered future. Although Arusha is now a far larger city, boasting a weekly newspaper *The Arusha Times* (available on the Web), a Rotary Club, and a new Rehabilitation Centre, much of it is still rundown and desperately poor. Recently (as Paul Theroux reports in *Dark Star Safari: Overland from Cairo to Capetown* [2003]) people spied a thief there and began to chase him—"Thief! Thief!" When they caught him they knocked him down and beat him to death.

The first miles we drive north through *shambas* of maize and bananas, but soon we come to rolling pasturelands dotted with cattle and goats, and eventually to arid soil covered with thorn scrub and scattered tussocks of grass: the land of the Masai. The bus stops now and then in the middle of nowhere to let a Masai man or woman in third class off or on—how do they know where they are when there are no visible landmarks? They wear large hoop earrings in a hole punched into the tops of their ears, and sometimes wooden plugs in their earholes; many carry long gourds, thin like themselves. The men are dressed in a single blanket, fastened at one shoulder, while the women, who wear more clothing, often sport large necklaces.

"You can always tell Masai from the other tall people who live around here—the WaArush, the WaKavia, the WaGogo," the artist tells me.

"You can tell because a Masai person ignores you unless you speak to him in Masai."

She and I have tea together at a hotel in Namanga, where the bus stops for an hour. This is the entrance to the Amboseli National Park, but there is an air of malaise in the town which seems to indicate that few people actually visit this park. Certainly there's no centre of biological research here.

About fifty miles south of Nairobi, we begin to spot game animals from the bus window—herds of giraffe, ostrich, hartebeest, impala, Thomson's gazelles, and zebra—amazingly tame to be so near a large city. They inhabit a vast savannah of grass where there are no trees, and the tallest vegetation, thorn bushes, are only eight feet or so high.

The bus stops in Nairobi at four o'clock, in a poor Indian district where Paul Theroux also arrives, forty-four years later, in 2001. He is unnerved by the swarms of people—hucksters, prostitutes, pickpockets, jostling youths, "urchins snatching and begging, as well as the blind, the leprous, the maimed." Single white women are followed by mobs of hungry children, threatening and pleading. Theroux, wearing second-hand clothes, no camera, and no obvious valuables, is relatively safe because there's no point in robbing him. Another author, Tanya Shaffer, writes in her recent book *Somebody's Heart Is Burning: A Tale of a Woman Wanderer in Africa,* that everyone she met in Nairobi, without exception, had been robbed. Even knowing this, she herself, purposely devoid of jewelry and a watch, had cash stolen from her pocket while shopping at a Nairobi market.

In contrast to Theroux and Shaffer, I step out of the bus into an empty, open square where I immediately hire a metered taxi to take me to the Norfolk Hotel. The streets of Nairobi are a shock after months away from sophisticated urban life because of the traffic—eight lanes of double-decker buses, lorries, and cars whiz along Delamere Avenue. The fabled Norfolk Hotel has, since its creation in 1904, been the Nairobi base of writers Ernest Hemingway, Karen Blixen, and Elspeth Huxley. In the early days it served as a gathering place for hunting safaris, the largest ever that of Theodore Roosevelt, who commandeered five hundred porters, each carrying sixty-pound loads. During the First World War, British troops marshalled here before marching west to the Lake Victoria region to battle the Germans from Tanganyika; then, for decades, settlers and visitors, many royal, partied and swapped spouses. In 1980 it will be partly destroyed by an Arab bomb, a chilling portent of future terrorism, although it will close for only two days before getting back to business.

The hotel is an enormous, rambling building which is, to my romantic eyes, disappointingly Europeanized; there are at least one hundred rooms and a huge dining area. A card propped up in my room reads, all in capital letters:

BEWARE OF THIEVES. GUARD YOUR MONEY. DON'T RELY ON LOCKS.
HAND VALUABLES INTO OFFICE FOR SAFE KEEPING.
YOU'VE BEEN WARNED.

Hardly a cordial welcome. No one is friendly except for the waiters who, as part of their job, greet me with a *"jambo."*

I already know from the letters of Dr. Leakey of the Coryndon Museum and Mr. Cowie, director of the Kenya National Parks, that there is no real possibility of studying giraffe using Nairobi as a base, but I go to visit them anyway—hope springing eternal. My effort is in vain. Mr. Cowie isn't in, a secretary tells me, and in any case is just off on a trip to England. When I visit the Coryndon Museum, Dr. Leakey is also away. I wander around the museum for a while, interested to see Leakey's fossil skull of Proconsul man and a variety of other African animals. In general, though, I'm depressed that there is no way for me to study giraffe from Nairobi any more than from Arusha.

As I return to the hotel, I cross a stream by way of a bridge over which a large sign proclaims in English and Swahili—"Bilharzia—do not wash or drink." In the stream, along with the snails which pass this debilitating and devastating disease on to people, five little African boys are splashing and shouting.

The next day I take a city bus on the route that comes nearest to the Nairobi National Park so that I can perhaps catch a glimpse of giraffe. The only way to actually visit the park and its giraffe, whom my colleague Bristol Foster will later study in detail, is by driving through it in a vehicle, which I don't have.

My seatmate on the bus introduces himself as Captain Fred Crystal, a retired South African (not interested in racial questions) who once farmed in Canada. I try to look out the window at the sights of Nairobi as we putter along, taking in and dropping off African customers, but find it difficult with his incessant chatter.

"Our sitting and talking together is quite proper," he assures me, who had certainly not questioned this. "I'm quite alone in the world and if I were thirty or forty years younger, we might be very interested in each other." He gives me a poke in the ribs, but I'm too polite to say that I certainly doubt this.

"I had a son, but he was killed by polio during the war," he continues. "Then I married a French woman who was always drunk and wouldn't try to smarten up because she said this was her fate. I've just divorced her." He goes on and on, but I stop listening to him, and just reply "Uh huh" and "Oh" at intervals as I look out the window. There are no giraffe visible inside the immense Park from the bus route.

On the way downtown again, Capt. Crystal gets off at the New Stanley Hotel. "If you want, we can have dinner together tonight," he says gallantly. "All the women at my hotel are faded English beauties trying to look younger than they are, which is why I'm taken with you."

"Oh," I say. "I'll phone you if I'm free," which I know I won't be and am not. I eat by myself at my hotel.

I've come to a decision about the future. There is no point fooling myself any longer that I'm going to find a way to study the giraffe of East Africa. Maybe if I had backing from some wealthy institution or funding agency (which will benefit virtually all later researchers), or a rich inheritance, I could figure out how to do this, but my funds are strictly limited if I hope to be able to pay my way back to Britain and then to Canada. My best bet is to return to Fleur de Lys, and continue my study of the giraffe there. In addition, I'll have a great deal of help there from Mr. Matthew in making a really good giraffe movie, the first ever produced.

That evening I consult the maps and tourist information I've gathered and decide at least to see as much of Africa inhabited by giraffe as I can. Rather than return to South Africa by ship, I'll carry on overland. I first consider the possibility of heading north into the dry region of Kenya where the *reticulata* giraffe hang out, perhaps the most handsome subspecies or race with their rich brown spots fitted closely together like paving stones, separated by a thin pale network or reticulation. These animals look so distinctive that they were thought for fifty years to belong to a different species than the rest of the giraffe further south and west with their blotchy spots.

*Reticulata* giraffe living in northern Kenya are of special interest to me because they exist in regions so arid that they may not have a chance to drink water for weeks or months on end. Does this mean they have evolved a more efficient cooling system than other giraffe? The variety of leaves available for their food is limited and adapted to desert conditions, so has their digestive system evolved to handle this? Many giraffe will not be able to congregate together with the limited food supply, so is their behaviour adapted to this reduced sociality?

Then there's the question of their spots. Why are these so different than those of other giraffe? Is it just a matter of chance? Or do they somehow camouflage the animals better? Are the angles of their reticulations somehow related to the angles of branches of the bushes among which they browse, making them hard for lions to detect? Or is their colouring somehow adapted

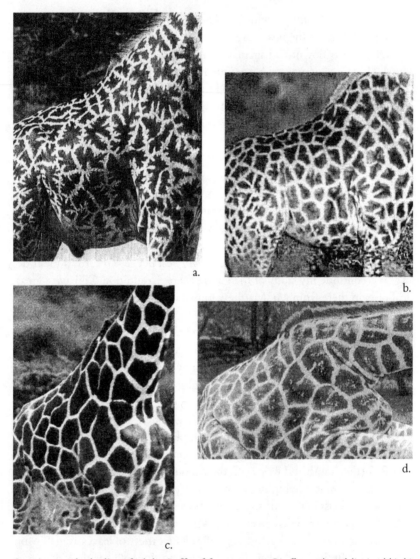

Spotting on the bodies of adult giraffe of four races: a. *Giraffa camelopardalis tippelskirchi* from Tanzania; b. *G.c. giraffa* from South Africa; c. *G.c. reticulata* from northern Kenya; d. *G.c. rothschildi* from Uganda.

to the strong equatorial sunlight? I would love to see *reticulata* in their native habitat to determine if these theories, suggested by biologists, make any sense.

However, there's no possibility of my doing this. Few people live in this desert area so there are few roads, let alone a reliable bus service. Even if I did buy a tent and camp out there by myself, there would be relatively few giraffe in any one place, given the limited vegetation, and I would have no vehicle to hunt for them. Nor would I have any way to access food and water for myself, nor to protect myself from possible harm from animals or men.

I decide instead to head west to the *rothschildi* giraffe in Uganda, a subspecies "discovered" in 1901 and named for Lionel Rothschild, who founded the Tring Museum in southern England. Old males are said to possess five horns rather than just the main ones, with two jutting out of the back of the skull which serve for the attachment of neck muscles, and one on top of the long nose, which presumably makes the skull a more effective weapon in male head-hitting battles. However, males of other races may also have such bony growths. The *rothschildi* subspecies is further puzzling because its spotting pattern shades into that of surrounding races; surely they interbreed? It has a much smaller range than the *tippelskirchi* giraffe with their wild variety of irregular spots, the only race I've seen so far in East Africa.

My travel folders indicate that I can board a steamboat on Lake Victoria and sail north part way around the lake into northwest Uganda to see the *rothschildi* giraffe, which seems like a good way to begin my return to the Transvaal. From there I can head south by boat along the west side of the lake and into Tanganyika again, where there is a train. I write to Martin English to tell him my likely plans, hoping he's still willing to let me drive with him south through Nyasaland; to Ian suggesting that we meet in England in the summer if he doesn't want to come to Africa; and to Mr. Matthew, to ensure that I'm still welcome if I arrive at Fleur de Lys, sometime in April, to stay for several more months, completing my report on giraffe and filming their activity.

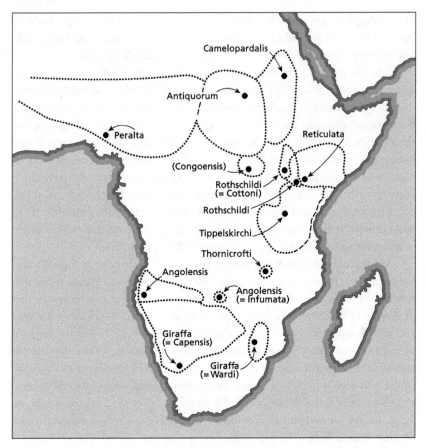

Maximum ranges of subspecies of giraffe, most now much reduced.
From *The Giraffe* by Anne Innis Dagg and J. Bristol Foster, 1976.

# 14

# *Heading South*

I nearly miss the second-class bus to Kisumu, a town on the northeast shore of Lake Victoria, because the hotel is slow in providing sandwiches and my taxi takes me by mistake to the first-class bus station; everyone seems determined that white people should travel the most expensive and luxurious way possible. His mistake is understandable, given that I am not only the sole white person in the small, crowded, second-class front portion of the other Kisumu bus, but the only person who speaks English. Before I can board it, I'm hailed by Capt. Crystal, who is taking the bus to Arusha.

"Why didn't you call me about dinner last night?" he demands.

"I met a friend and couldn't get away," I lie.

He gives me his address on a bit of paper and asks for mine, although I never hear from him.

I clamber into the bus to join an Indian couple with a small boy, a large African woman with her baby whose face is covered with flies, and three Indian men; the back, third-class section is crowded with Africans. One of the Indian men, an old Sikh in a turban, shows me what seems to be his business card, which says in the corner that he is a piano tuner. From his gestures I assume that he also plays, and possibly tunes, other instruments, including organs and violins.

The road to Kisumu is travelled not only by a few buses and other vehicles, but by Kikuyu, especially women, bent over hauling huge loads of firewood and other goods on their backs, taking much of the weight on tump lines pulled taut across their foreheads. We pass fields of corn and bananas, and a man ploughing a few acres with a team of four oxen. Insects disturbed by the plough are snatched up by large storks stalking along behind it. In this

White Highland area, only Europeans are allowed by law to raise cash crops; Indians are restricted to owning shops, and Africans to work as farm labourers.

Soon we come to the edge of the Rift Valley, with a magnificent view ignored by everyone but me. As we wind down into the valley, the vegetation changes from lush green prosperity to a dusty bleakness, broken occasionally by well-kept European farms where plump Holstein and Ayrshire cows graze in fenced fields.

Villages are not the casual sprawl of dwellings I've seen up until now, but five hundred or more neat, round, thatched huts in rows, each with a small garden, grass plots, and a back outhouse. Their army-encampment appearance has been mandated by the British, who have been trying to keep a lid on the recent Mau Mau (pronounced to rhyme with "now-now") revolt of members of the Kikuyu tribe against British rule. Such orderly villages are easier to defend in the event of night attacks by Mau Mau. Fortunately, violence has died down, so I feel in no personal danger; the British government combatted the Mau Mau without mercy by executing hundreds of them, as well as people who consorted with them or who carried illegal weapons.

I've read a great deal about the Mau Mau and their tactics of killing with knives and swords both white people and thousands of their fellow Kikuyu who don't support their campaign of violence. These victims had often refused to take part in the Mau Mau Oath of Allegiance, which involved crawling through an arch of banana leaves seven times while muttering an incantation—seven times representing the seven orifices of the human body. Being a zoologist, I worry about the seven orifices. A mouth, two ears, rectum, urinary tract, yes. But does the nose count for two when the nasal pathway leads to a common tube? What about the eyes? Are these really orifices? And what about women's reproductive tracts which are involved in menstruation? If men have seven orifices, then women must have eight. Yet women are certainly involved in the revolt. There is no one with whom I feel free to discuss these pressing questions.

As the bus trundles along the road, the Sikh piano tuner across the aisle catches my eye when we pass a house that has a piano. "Piano," he says loudly, pointing to the house with a big smile, at which I nod and smile back. Once he says "organ" at a large mansion, which elicits more enthusiasm from me, and once he indicates a school with *five* pianos, an attribute that makes us both observe the building with suitable wonderment. We are getting along so well that the Sikh is annoyed when a small, elderly, African man who speaks English boards the bus and squeezes into a corner seat.

"You are from England?" this newcomer asks me in English.

"No, Canada."

"Canada is cold, isn't it," he states, looking around aggressively at the other passengers who have no idea what he is saying. The African rolls his eyes at their stupidity and repeats his observation in some native language that they understand. Some of them nod vaguely.

"This woman is from Canada," he says in English, then shrugs contemptuously again at the blank looks of the others before translating his sentence for them. At intervals he cowers down in his corner to take a swig from a brandy bottle he carries in his coat pocket.

I'm the only one with sandwiches on the bus, and when I offer him and the others one (the Sikh has already declined), he is so touched that he offers me a swig of his brandy in return, which I refuse. In the excitement the African slips off his seat onto the floor of the bus.

"I need brandy to settle my stomach on bus trips," he explains to me as he picks himself up unsteadily.

The Sikh looks at me with furrowed eyebrows to express his horror at the African's drinking. I pretend not to know what he means.

After several hours the driver stops the bus at Nakuru for a toilet break. He indicates that he will drop me at the hotel if I wish, but I refuse his kind offer on principle. Most of the passengers get off the bus and hurry to one of two double outhouses, one labelled for African men and women and the other for Asian men and women. This may be another reason why "Europeans" don't often ride second class. The men relieve themselves on the outside of the huts, so I wait until they have returned to the bus before deciding to become an Asian woman. (Do I fancy that Asian women are more fastidious than African women? Or that I'm somehow more closely aligned with them? Or reason hopefully that there are fewer of them?) I approach the Asian hut, retreat briefly when I come across a man defecating beside it, then pull back the wooden door to find a hole in the ground, surrounded by evidence that many women have missed it. No wonder the smell and flies are so overpowering. I'm glad I'm not a real Asian woman, because my trousers are easier to clutch above the mess than skirts or saris. Recent travellers' accounts of their experiences in Kenya report that these revolting toilets on bus routes remain revolting to this day.

When I return to my seat a little boy is being cleaned up after defecating into a potty which is pushed, uncovered, under his parents' seat. Because of the smell I drink the Pepsi-Cola the Sikh has bought me outside, displaying obvious relish for his sake. At a later small *duka,* owned by an Indian, the small African man offers to buy me a beer or a brandy, the Sikh a second soft drink, and a man I haven't seen before anything I want to quench my thirst. By now I'm aware that outhouses are rare, so although the men can continue to climb out of the bus and urinate with their backs to the road, the women have to contain themselves. I refuse all the men's generosity despite the heat, knowing what urination entails. How sad that I alone am treated so generously, when I probably have more money than any of the other people on the bus.

At a village called Amassi, the bus stops at a native market where hundreds of people mill around, some women walking with huge jugs balanced on their heads. An Indian inspector drives up to check all of our tickets. When we finally arrive at Kisumu, about five, this man is waiting to drive me and my luggage to the Kisumu Hotel.

"I shouldn't have special service," I object, although I am delighted by his offer, since it has begun to rain.

"It's all in the service," he explains lightly, although it obviously isn't because all the other passengers have to walk or take a taxi to reach their destinations. Again I'm aware of the consternation local people feel on seeing a European woman travelling second class.

Kisumu has between five hundred and six hundred Europeans but, although the hotel is full of British people keeping to themselves (most dine at separate small tables), few are in evidence when I walk through town on the way to the dock the next morning. Lake Victoria, stretching out before me, is immense—250 miles long, and the second largest body of fresh water in the world after Canada's Lake Superior.

The station master at the steamship ticket office is very friendly. "It's nice to see a Canadian who isn't a dowdy missionary," he compliments me, although I'm dusty and sweaty from the heat.

"I want to go to Uganda," I say, "and then on to Mwanza at the south end of Lake Victoria. Is there a boat leaving today?"

"Boats don't go that often," he replies. "That'll take more than two weeks. If you want to catch the train south from Mwanza, you should take the boat sailing clockwise around the lake. There's one leaving here tomorrow evening."

I don't bother to explain to him that if I do that, there's no hope of seeing *rothschildi* giraffe in Uganda. But if it takes over two weeks to reach Mwanza the long way around, I'll miss a great deal of photography time at Fleur de Lys. I'm not sure what to do. "I'll think about it and come back," I say.

As I'm walking from the wharf to the hotel, an engineer driving past stops to give me a lift into town; he can't bear to see a white woman out in the sweltering sun.

Market day at Kisumu, Kenya.

"It's too hot to walk. Do you know Kisumu?" he asks. "I haven't seen you before."

"I've just come to catch a boat," I say.

"Let me show you around then. Come to the house first for a drink to cool off." Seeing me about to refuse, he laughs and adds, "I'll introduce you to my wife." I accept his invitation with pleasure.

Ray and Jack Somerville live far better in Kenya than they would in England where they come from, with a smart new bungalow, a new car, and servants to do all the housework.

"We miss our kids, though. They're both at school in Nakuru, where it isn't so hot," Ray says.

The three of us walk to the native market, where African women and a few men sit behind small heaps of vegetables piled on cloths on the ground, most chatting to each other.

"The government has a ceiling price for all produce, most of it native, and controls the price and quality of milk and meat," Jack says.

"The shops are almost all run by Indians," Ray says, "so the government has built ten small shops over there which can only be rented by Africans." She waves toward a row of small buildings in the distance.

We drive around town, past a tame herd of impala wandering along a street, and by the main buildings, including two churches, one Anglican and the other Roman Catholic, and the mosque. Sweating African men are lugging lumber about on building sites, constructing a number of new houses.

"Everyone who works for the railway gets a free house," Ray says. Then, changing the subject as we approach the lake, "On some nights, hippos come right up here from the lake to feed."

We visit the two clubs in town for a drink (in my case an orange squash), a prime pastime for Europeans in Africa. The Nyasa Club, which costs £10 to join and thirty shillings a month, has a swimming pool, tennis courts, a golf course, and a clubhouse with bar. The less ostentatious Railway Club, for railroad personnel, has only a clubhouse with a bar, billiards, and badminton. I tell the Somervilles about my desire to visit Uganda so that I can see the giraffe there. Not just any giraffe, of course, but *rothschildi* giraffe.

"It won't be easy," says Jack. "There are so many people in Uganda that giraffe have been pushed far back from Lake Victoria, if they ever even lived near there. There are some in Kidepo National Park, and some near Murchison Falls and in the far north. I didn't know there were different kinds of giraffe, though."

"Would buses go there?" I ask.

"None would go very near giraffe. You'd have to rent a car or take a taxi."

"It sounds expensive."

"It would be."

After mulling this information around for a while, I decide to forget about the *rothschildi* giraffe and instead book passage on the boat going south. Jack drives me to the dock so I can do this, then back to my hotel where I write to Martin about this change in plan, hoping the letter will arrive before I do.

The next morning I fall into conversation at the hotel with a twenty-year-old called Dan Confait, an Afrikaner from South Africa in charge of transport for the Prison Department.

"You should have been here for the Mau Mau raids," he tells me proudly. "I shot seventy Mau Mau, including General China. Do you want to see the prison? I can drive you there when I go this afternoon."

Before I can answer, Ray and Jack arrive with a boat-builder friend, John Robinson, who will be travelling on the same boat, the *Usinga*, as me. The four of us have a drink at the hotel, another at the Railway Club, and then a lovely lunch at the Somervilles' home, prepared and served by their "boys." Entertaining is as effortless in East Africa as it is in South Africa.

When I return to the hotel after lunch, Dan is just leaving for the prison.

"Do you want to come along?" he asks again. "It won't take long. I have to make a delivery."

"Sure," I say. I have nothing better to do.

We drive seven miles in a Land Rover to a huge area surrounded by a fence and a twenty-foot-deep moat. Inside there are tidy buildings to house the warden, guards, and inmates, with roads and paths lined with rocks painted white. In one corner prisoners are making bricks; in another area are small neat gardens of sisal, vegetables, bananas, and sugar cane.

"Women aren't allowed in the prison," Dan tells me. "There's too much danger they'd be taken hostage and raped. We can go into the hospital, though." I assume the prisoners there are too sick and feeble for violence. It's difficult to say. Behind the locked door African men are lolling on iron cots, most appearing benign but some looking fierce.

"That boy sitting on the ground's a Mau Mau," Dan remarks, pointing to a small, nondescript man dressed in white leaning against the wall.

"Let's not loiter," I say.

On the way back to the Land Rover, Dan predicts that the Mau Mau rebellion isn't over. "It'll probably be revived in a couple of years. I hear there's a huge training school up north where natives are taught to shoot to kill. But I'll be ready for them." He puts his arms up as if to aim a rifle and pretends to kill everyone in an arc in front of him.

"I don't mind natives," he goes on amiably, "but they don't like me because I make them do what I tell them." On the way back to the hotel we stop briefly at the Remand Prison, where groups of convicts are squatting in line so that they can be counted before entering the prison.

Back at the hotel, we have tea. "Girls like me," he says. "The only trouble is there are only three single girls in Kisumu."

"That's hard on you," I say, but he misses my sarcasm.

"There are some nurses, of course, but they're about twenty-five. That's far too old." I don't tell him I'm now twenty-four.

After tea, we stroll down to the beach. "Floating islands land at Kisumu about once a month," Dan tells me. "They're full of snakes."

"That doesn't make sense," I say, but there is indeed a large, raft-like clump of roots and vegetation, about four feet thick, touching the sandy beach. When I leap onto it to explore, I land beside the moult of a big puff adder. I promptly leap off again.

"A snake!" I exclaim.

"That's what I said," retorts Dan. "Once there was a thirty-seven-foot python."

This I don't believe and I begin scoffing again, which Dan doesn't seem to mind.

Back at the hotel Dan gives me a picture of himself with crew-cut hair, which makes him resemble a convict. I'm about to say this when I realize he thinks he looks good.

"Thanks," I say.

After dinner, the Somervilles drive John and me to the Usinga where we have drinks on board before it sails at 8:30 p.m. We watch the crew tie ropes so that the boat can turn itself around by pulling on them before setting out for the south.

The *Usinga* is the size of a small ferry. In first class, where I'm booked, there are fifteen cabins, each with two beds and hot and cold running water. Second-class cabins, mostly inhabited by Asians, have four beds (two bunks) per cabin while third class, at the back of the boat, is merely space on deck under large tarpaulins. This is where about three hundred Africans sit or lie among their belongings. They pay only a few shillings to travel this way, but at each stop they have to troop onto the dock with their possessions—gourds, spears, bicycles, pots, and bundles—so that the crew can unload and load cargo in the hold. The boats planned to carry passengers as well as cargo below deck, but no one wanted to travel this way, much preferring even cold wind and rain in the open. The ferry has a lounge and two small saloons serving good meals, but most Asians and Africans eat food that they've brought with them, either because it's cheaper or because it's more to their taste. The ferries still exist today because people want to travel from one port to another, but the boats are in such bad repair that the service is too irregular to be of much use.

After breakfast it's too cold and rainy to sit outside if you don't have to, so I retreat to the lounge where a middle-aged woman and two young men are camped. I talk to Ivan, a nineteen-year-old Mennonite from Maryland going to the next stop, Musoma, in Tanganyika, to build a Mennonite church and so evade National Service in the United States. The shoreline to our left is now Tanganyika rather than Kenya, but there are no immigration problems.

"If I work in Africa for three years, I don't have to go into the Army at home for my two years of National Service," he says.

"Are you a conscientious objector?"

"Yes. Mennonites don't believe in killing people."

"Few people do," I retort. "What would you have done if Nazis had moved into your town and killed anyone they didn't like? They would have continued doing that in Europe if the Allies hadn't fought them."

"I would have prayed to God for help," he says earnestly. "Prayer is more important to a nation than weapons. God would have sent a plague or locusts or something to Germany. Remember the walls of Jericho?"

"You must be very religious to believe that," I say. He doesn't seem to mind my saying this.

"My father's head of a mission in Maryland. We don't have a church yet because there are only about 120 Mennonites there, mostly tenant farmers who don't dress well enough to go to other churches. Mennonites *are* very religious," he says. "I don't drink, or smoke, or dance, or wear jewelry, or play cards—except Flinch, now and then.

"Flinch?"

"Flinch isn't the same as ordinary cards."

"What about your fancy timepiece?" He wears a large watch on a jazzy, silver expandable bracelet.

"I need a watch to tell the time," he says defensively, "but I guess the strap is a bit gaudy." He wrinkles his forehead while admiring it. "We don't go to movies, either, except I snuck out once to see The Ten Commandments."

"You devil, you."

"I know I shouldn't have," he says, oblivious to my teasing, "but I don't think it hurt me."

"Do you believe in evolution?" I ask.

"Of course not."

"Do you date girls?"

"Of course. But not usually Mennonite girls. They have to wear shapeless, ugly, two-piece dresses." Ivan himself wears a snappy yellow sports shirt. "I had a job as an IBM operator in the States and earned lots of money, but in Musoma I'll only get $500 a year."

When Ivan and the Mennonite woman leave to pack before we dock at Musoma, I chat to Norman Horsley, a lad from Mwanza who says he shoots elephants, and whom the Mennonite woman was trying to convert to Christianity.

"What a bore," he says. "She's been in Musoma for twenty years and she thinks of nothing but religion. I wonder if it's true that American missionaries hope to take over Tanganyika after the British are thrown out? Heaven help us if they do!"

After lunch I watch the myriad of third-class passengers straggling down the gangplank with their belongings and then the large cranes unloading cargo from the hold. Convicts, wearing loose, white shorts and white shirts with broad green stripes down the front and back to denote their status, carry smaller loads ashore.

Then, with an hour left before we sail, I walk about half a mile into town, the only white person abroad as far as I can tell. Some Africans sit in front of their neat, square, thatched houses while others stroll about and chat. Most just stare at me in a friendly way, but a few are willing to exchange "jambos." I stroll behind a little boy with a bundle on his head, walking in turn behind his father. He looks back to stare at me with huge eyes every few minutes, but when I smile at him he rushes ahead in confusion, bumping into his father's legs. Finally he grins back shyly, then ducks his head. Near the quay on my return to the *Usinga,* a somewhat larger boy, an Arab this time, asks if I want to hire him as my boy.

After the boat sails, I chat with Norman again over dinner and during the evening as we sprawl on deck chairs with our feet up, watching rats and mice scamper on deck among the crowded, sleeping bodies. A few Asian people are seasick over the side of the *Usinga,* although the lake isn't very rough. A large cockroach somehow ends up in my empty coke bottle.

"My job is to lay down telegraph wire through tsetse bush areas. I have 180 boys working for me who have to be tested regularly for sleeping sickness."

"That's a lot of men," I say.

"You bet. Whenever I get a new bunch of workers, I have to show them who's boss at the very beginning. I've been fined for hitting a boy, but you have to keep them in line. Otherwise they won't respect you and you can't keep order."

"You live in the bush?" I ask.

"Yes, when I'm working. I shoot meat for my boys once a week. I can get a licence for two elephants a year which I shoot with my .404; tusks sell for £1 for a pound of ivory, so I make good money from them. I shot a crocodile once, too, and sold the belly skin for £10."

"Are there many animals where you live?" I ask.

"Not too many outside the reserves. The elephants know where the reserve boundary is; they'll run over the border when they're hunted, and then relax and have a drink. Pretty smart."

"Any snakes?"

"Lots. One python knocked down a teacher walking after dark on the beach near my house at Mwanza. It bit her on the ankle before two boys rescued her. We also had a mamba in the bathroom of our guest house in Dar es Salaam."

"Will you stay in Tanganyika if the British leave?" I ask.

"I don't know. I'd like to go to Canada, because there's not much future here. Shooting grizzlies would suit me. I think there'll be a blow-up, like the Mau Mau in Kenya. Probably the Chagga tribe will start it, because they're smart and well-organized." As we converse, an African employee brings me my shoes, which he has shined. He bows in thanks when I give him a shilling.

From the ferry, in the early dawn, Mwanza looks like a dense conglomeration of elephants, with immense grey boulders, some larger than houses, jumbled along the shore. When we debark, Norman takes me to meet his

family and his girlfriend, Glenda. We spend the morning at his house playing records and chatting, and the hot afternoon strolling along streets lined with
flowering trees. Mwanza is hedged in by the lake, thick bush, the huge boulder outcroppings, and steep hills. We wander through a quiet cemetery where
the gravestones tell of a fierce battle between Germans and the Belgians and
British on June 23, 1915. From the top of a hill, sitting on the ground among
brazen lizards and timid dassies, we have a magnificent view north over Lake
Victoria. After tea at Norman's flat, his mother gives me a sandwich she has
kindly packed for me and we drive to the train which will take me south to
Tabora and east to Itigi.

The start of my land trip south is jolly, with John Robinson, Phil Armstrong, and me joined by four of Phil's friends, who are travelling only one
short stop to continue their drinking and partying; John and Phil are taking
the train all the way to Dar es Salaam for their two-week local leave. When I
suddenly remember that I left my raincoat in my cabin on the *Usinga,* Phil
comes to my rescue.

"I work for the railroad," he says. "I'll send a telegram to Kisumu to hold
your coat. When I get back after my leave, I'll mail it to you."

I give him my address at Fleur de Lys, amazed at his kindness when I'm
almost a stranger to him.

At the third stop, I talk on the platform to a raggedly dressed European,
who turns out to be the engine driver. He invites me to visit in the engine cabin
which I do, meeting there a handsome Sikh engineer taking local leave and
an African shovelling coal onto the engine fire. As we chug south I lean out
the doorway to wave at African farmers in small *shambas* growing rice, corn,
potatoes, and cassava, with hedges of sisal. Occasionally there are fish traps set
up in sluggish streams, and scattered herds of cattle and goats.

When it grows dark about six o'clock, the engineer sets the engine at a constant speed of twenty-five mph, the maximum allowed, and we sit down to
chat near the middle of the cab, since rain is now splashing through the open
cabin sides. I'm the only one who looks nervously down the track every few
moments, afraid that we're going to bump into something.

To entertain us, the Sikh takes off his turban so we can see his hair fastened in a little bun on the crown of his head. He winds up his beard each day
around a piece of string, then ties that on top of this head, too.

"It takes me an hour to fix my hair every day, and five minutes to wrap
my turban," he explains as he winds the long cloth around and around his
head. On his upper arm he wears a bangle, with a small box containing a religious icon.

"In Canada it's women who spend time on their hair," I laugh.

"Can you get me into Canada?" he asks. "There isn't going to be much of a future here in Tanganyika."

I realize that I know nothing about immigration to Canada. Is anyone allowed in besides DPs—persons from European countries displaced by the war? I don't remember seeing Sikhs in Canada. Does one apply somewhere, or write to Ottawa? I'm hopeless as a contact for him. I feel badly to be so useless.

The white engineer invites me to have bacon and eggs with him when he gets off duty after midnight, but I refuse; by then I'll be sound asleep, the only person in my compartment. At the next stop, I return to first class and go to dinner with Phil and John, who have been wondering where on earth I was since I wasn't anywhere in the carriages.

After breakfast we arrive at Tabora on the main east/west railroad built by the Germans a hundred years ago to ship raw resources to Dar es Salaam on the coast. Tabora was well-known as an important slave trade centre, with large houses and gardens owned by wealthy Arabs. I begin to walk into the town centre, but find it too far to reach in the fifteen minutes available. When I return to the train, a jovial English woman is moving into my compartment, laughing loudly.

"I saw a snake on the platform outside and I called a boy to get it," she chuckles. "A boy came, but he brought a bow and arrow. By then the snake had slipped into the bush."

Miss Knight is a Girl Guide organizer, whose job is to establish and monitor Girl Guide troops around the country.

"We have about three thousand Guides and Brownies in Tanganyika," she tells me proudly. "Not many in the bush, but lots in the cities and at schools connected with missions."

"Do they have the same structure as Guides in Canada?" I ask as a former keen Brownie, Guide, and Brownie Pack Helper.

"Yes," she answers. "We've had the Guide Handbook translated into Swahili. The tests are almost the same as those for English girls except, of course, Brownies wash up the cooking pot instead of the tea things."

"Do girls of all races belong?"

"Of course. More of the Asian girls become First Class Guides, but this is because they're more likely than African girls to have European leaders, who push harder than African ones." Miss Knight would be proud to know that in 2000 there will be fifteen thousand Tanzanian Girl Guides. Today, many Africans would see this British Girl Guide legacy ambivalently, partly as a source of pride but at the same time as a remnant of cultural colonialism. Even though British rule had long ended, African girls are still subject to mildly modified British traditions rather than to their own cultural heritage.

The scenery that creeps by as we chat is monotonous, with strips of grass twenty yards wide on either side of the tracks and then thick thorn bushes extending to the horizon—probably Norman's tsetse fly bush because there's little evidence of human life or of livestock. The few Africans live in small houses clustered around each train station.

In 2001 Paul Theroux, on this same train, passes by scenery which sounds unchanged. The train, however, is much worse. The dining car is filthy, serving food he'd rather not eat. Men too lazy to go to the toilet use the sinks in their compartments as urinals—or maybe they do this because the toilets themselves are vile, encrusted with dried urine and feces. Late one afternoon he reports that three drunken teenagers trap a girl in the toilet but, before they can rape her, her screams bring help from the conductor.

In the afternoon over four decades earlier, on our far cleaner train, we arrive at Itigi where I'll stay until I board a bus the next day heading south for Mbeya. Miss Knight, Phil, and John all escort me to a "Superior Class" Dak Bungalow, one of many built to accommodate travelling Europeans in towns where there are no hotels. We pass a jet black African stalking along the road with huge white discs in his ears, a loose white toga over his body, and a bow and arrow in one hand. The road surface is so hot that our shoes stick to the melting tar; my companions quickly wish me luck at the bungalow so they can return as soon as possible to the cooler train.

It's a relief to be settled in one spot for a day. I'm the only overnight guest at the bungalow, so after dinner I'll have the whole house to myself: a large bedroom with three beds for women, another for men, each with a lavatory, a screened verandah, and a large dining room for residents and local Europeans. There are at least three African "houseboys" to look after any guests (me), and an obsequious Indian man in charge.

The house is full of flies, which I swat with a folded newspaper. As I'm finishing this self-appointed task, a white couple with two children drop by for drinks.

"Hello," I say putting my weapon aside, delighted to have company. They ignore me. Maybe they haven't heard me?

"Hi! How are you?"

"Hello," they answer coldly. They leave the door open, which attracts more flies into the dining room, and settle down at a small table on the verandah with their backs to me. When they finish their drinks, they leave without saying goodbye.

"The Europeans aren't very friendly," the Indian man says, standing by the dining table where I am seated, waiting for dinner. "There are only two European and nine Asian families in town, and of course many Africans."

"And flies," I laugh.

An African man brings the dinner which has been prepared for me alone. Two ants and a fly float in the bowl of soup, but the waiter is unperturbed when I call this to his attention. He takes my fork to fish them out.

After dinner I bring *The Journals of André Gide* (I finished *Crime and Punishment* long ago) from my room to read, but the Indian is determined to keep me company. When I settle into a chair, assuming that he will leave, he instead comes behind me and begins to caress my head.

"You have lovely hair," he says, drawing it through his fingers.

I jump to my feet to escape this attention. I realize suddenly from the silence that the African help have gone home and I'm all alone with this man.

"It's very lonely here," he says. "I don't have any family to keep me company."

"I must go to bed now," I say firmly. "Good night." I march into the bedroom and close the door, wedging a chair under the door knob to prevent the man from entering, although he probably has no intention of trying to do so. I lie awake for almost an hour, worrying about being vulnerable in the house, but eventually fall asleep.

The next day after lunch, which consists of a sandwich complete with insect parts that crunch in my teeth as I chew, I pack and board the bus for Mbeya. I'm excited to be travelling on the Great North Road of Africa, but also dreading the overnight trip because I can never sleep sitting up. Again, the small front part of the vehicle with cushioned seats is second class; the larger, third-class section at the back, with wooden benches lining the sides, is crowded with Africans, hens, and an assortment of bulky parcels.

My seatmates include a man from Goa and an Indian.

"If I'd known you were in Itigi last night you could have come to a party," the Indian man exclaims to me. "It was so much fun that it didn't break up until eleven o'clock!"

The rest of the second-class passengers are all African men with the exception of an old Muslim woman, who sits in a back corner with a cloth over her head, coughing consumptively. The men, who speak to each other in English, pass around copies of *True Detective* and *Film Story* to read. One says "Jesuschrise," when he swears, another "See you later, alligator" when someone leaves the bus. When a jackal crosses the road in front of our lumbering vehicle, the men all turn to me to make sure that I see this example of African wildlife.

The road isn't like my idea of a Great North Road at all. In fact, it's only cleared earth, with grass sometimes growing down the middle. Every ten miles, the surrounding bush is interrupted by a clearing where the workers who maintain each section of the route live. We pass only two vehicles before dark, both lorries broken down by the side of the road. We give one two tires and the other a new axle. I realize now that we're travelling in a convoy of three "railway" vehicles, the others being another bus and a lorry. At midnight we stop because the lorry has broken down; by 1:30 a.m. the resident mechanic of the convoy has fixed it up again so we can continue to Kiyombe for a fill-up of petrol.

Here a missionary from the Catholic White Father Society, dressed in a long white robe, joins the bus and settles beside me so that we can talk as we bump along through the dark; the bus lights stay on all night, so it's difficult to doze.

"I've just returned from my leave in Canada," he tells me in a French-Canadian accent. "I had to recuperate from having fever, bilharzia, and stomach ulcers."

"It must be a hard life here," I say, thinking of the old woman coughing out germs behind me.

"I've been lucky," he replies. "As well as malaria there's lots of leprosy, sleeping sickness, and elephantiasis around. When I ride my motorcycle between villages, I wear nets over my head and hands, and two pairs of socks with newspapers between them on my feet so the tsetse flies can't bite me. I have a house at each village to sleep in, but if the lock on the door is smashed when I arrive at one of them, I won't sleep there in case I catch a bug from the man who broke in."

"You don't even do this for money," I say, looking at his strange garb.

"Of course not. It's worthwhile, though," he insists. "I make about one hundred converts to Christianity a year, even though there aren't many people around Kiyomba. Down near Mbeya, business is better." He is spending his life "saving" Africans, but from his comments, he doesn't seem to like them much.

At breakfast time we arrive at Chunya, a former gold mining centre now largely a ghost town, where I buy a sandwich to eat. Here three fat white men climb on board the bus for the short fifty-mile trip to Mbeya. Except the trip isn't short. I've been thinking of myself romantically as on the Great North Highway unfolding the length of Africa, but that is before it begins to rain. As the bus laboriously climbs up into the southern highlands of Tanganyika, the road becomes muddy as we pass plantations of pine and eucalyptus trees, and head further up into a rainforest jungle. The rear end of the vehicle begins to slither, several times ricocheting off one of the mud banks lining the road. Once we stop behind a row of lorries stuck in the mud which gradually free themselves, one by one, to lumber on. We carry on, slowly too, until ten miles from Mbeya.

Then, while trying to pass a lorry mired in mud on the narrow road, we become stuck ourselves. The African driver guns the motor to free our tires, but this makes us slide into the gulley on our right, the bus's side leaning against the right bank. The driver drops his shoulders in despair, opening the doors so his passengers can clamber off the bus into the rain to lighten its load. The European men stand apart while the African men push on the back and side of the bus to get it back onto the road. I start to push too, but one of the Africans offers me his raincoat and motions me to back off. This isn't women's work.

"Jesuschrise," an African curses as he strains.

The men labour and push and yell in the rain for an hour. By then the bus has edged backward and is close beside a second stuck lorry. Another bus and five more large lorries are now lined up behind us with no room to get by.

"Jesuschrise, Jesuschrise!"

Finally, men tie a rope to the back axle of the bus, and by tugging and pushing it sideways, we regain the road. The driver carefully steers the vehicle forward a hundred yards, then stops so that we passengers can follow through the muck and climb aboard. We sit tensely as the bus struggles to the top of the hill, then relax on realizing that the road winds downhill the rest of the way to Mbeya. It has taken us four hours to cover the fifty miles, and I've been sitting on the bus, without sleep and with little food, for three hundred miles and twenty hours.

Mbeya is a lovely, English-type town with small, red-roofed houses set among eucalyptus and conifer trees, its prosperity based on surrounding coffee plantations. But I'm not sure what to do here. I've missed the second-class bus to Nyasaland and don't know if Martin English is going to meet me, and if so where, because the post office is closed and I can't get the letter from him I'm supposed to pick up there. The next bus south, which leaves in half an hour, looks far more trying even than the one I've just left because it's all third class, and is already packed with Africans sitting on wooden benches facing each other with chickens and bundles stuffed into any extra space. There is no hotel before Lilongwe, five hundred miles south, and although there are a few Dak Bungalows, guests have to bring their own food, which I don't have, to stay in them.

I'm standing on the main dirt road beside the bus, looking about in despair when, to my joy and surprise, up drives Martin in a Ranch Waggon.

"You'd never manage on the bus," he says categorically, and opens the back of his vehicle so he can stow my suitcase and knapsack there.

# 15

## *Mbeya to Umtali*

I'm thrilled to join up with Martin—someone to chat with and who has a car! We eat lunch at the Mbeya Hotel, regaling each other with our adventures since we parted on the *Kenya Castle* months earlier. Mine are the more exciting.

"I had to come north to buy soap samples in Tanganyika," Martin tells me, "so I timed my peregrinations to fit in with yours." He still loves big words.

"That's super," I say.

"You won't find any giraffe in Nyasaland, though," he warns.

"I know. There are giraffe to the north and giraffe to the south, but no giraffe through the middle of Africa."

"A discontinuous distribution."

"Exactly! The vegetation in central Africa is too dense—like the jungle in the Belgian Congo. Sometime in the past the vegetation changed there to a type that giraffe didn't like. The northern animals became separated from the southern and different races evolved in both areas. I'm not sure when."

We pick up my mail at the post office which includes the letter from Martin. There is one from Mr. Matthew, who writes that he plans to be in Southern Rhodesia on business about the time that I'll arrive there, and wonders if I would like to take the train to Umtali and from there tour around a bit with him before driving back to Fleur de Lys. YES! Martin waits while I mail off a postcard to Mr. Matthew, agreeing with this wonderful plan. I ask him to write me care of the main post office in Salisbury, the capital of Southern Rhodesia.

As we set off south for Nyasaland, a white Southern Rhodesian lad asks for a lift, to which Martin agrees. John is returning to the family farm after travelling for a year to Singapore, Greece, Italy, France, and England. I expect

some interesting insight into these countries but am disappointed; rather, he
tells us how he keeps his shirts looking ironed while travelling—by packing
carefully.

We spend the night at Tunduma, still in Tanganyika, at a quiet inn surrounded by gloriously large trees, paying twenty shillings each for a room
and board. Tame Sikh monkeys, who hate women, and chic black and white
colobus monkeys hang about the buildings, hoping for food. One of the latter, a doleful expression on his face, teases two cats.

The three other guests and ourselves eat dinner together around a large
table with the owner, an elderly English woman; there is no electricity, so the
room is lit by pressure lamps, which give a homey atmosphere. One guest
is a Canadian called Joe, from Stratford, Ontario, who lives with his parents
in Kenya. He's an apprentice lorry driver who has been waiting a week,
apparently in vain, for his lorry to return from the south; for some reason,
he wasn't allowed over the border into Northern Rhodesia.

"But *Africans* drive lorries, not Europeans," Martin objects.

"That doesn't mean I can't too," says Joe, who doesn't seem to appreciate Martin's point—that Europeans are superior to natives and, because of this,
should never do native jobs. Joe has no change of clothes and only one shilling;
Martin gives him money so he can buy food and hitchhike back to Nairobi.
Martin never gives money to destitute Africans, which doesn't strike me at the
time.

Another guest, also named John like our hitchhiker, agrees to give our
John a lift in the morning. Our John is so delighted to be nearly home that
he can't stop chattering about himself.

"I'm anti-social," John II breaks in at last. "If you're going to talk this
much tomorrow I'll lock you in the boot." John I, crestfallen, immediately
shuts down for the evening.

Next morning the road south from Tunduma meanders randomly
through Nyasaland and Northern Rhodesia, not that it matters when there are
fewer than a dozen cars a day trailing along it. At the Customs Office of the
Federation of the Rhodesias and Nyasaland at Fort Hill, we're greeted so
cheerily by a white man that we wonder why hitchhiker Joe had problems
here. All through Nyasaland we will have clear sailing (if this metaphor can
apply to excessively bumpy roads), with no officials ever flagging us down.
Today, an itinerant white person is stopped at dozens of roadblocks, where uni-
formed Africans with guns examine passports and visas with care at each,
with at least a few hunting for flaws to be rectified by a handsome tip.

We visit two tiny shops in the next villages we come to so Martin can check
out their Lever Brothers supplies of soap.

"Do you get all the soap you want to sell? Enough Sunlight soap? Lifebuoy?" Martin demands of the first owner. Part of his job is to check up on the representatives for Levers Brothers in each area of Nyasaland. Martin can communicate in primitive Gujarati if necessary when he speaks to shop owners, but English is easier.

"I could use another case of Lifebuoy," the Indian replies politely.

"I'll check into it," say Martin, making a note to do so. "Can I buy an aspirin?" he asks, because he has a headache.

"I'll *give* you one," says the owner. "You can give me a free tablet of soap sometime in return." He laughs, but Martin seems not to be amused.

Several nattily dressed African men are hanging out by the *duka.* One, dressed in a suit, cowboy hat, and dark glasses with white rims, returns my smile.

"They're miners back from working in Mozambique," Martin says. "They purchase jazzy clothes to impress the women, and fancy bikes and trunks." He points out a flashy trunk beside the store, ornamented with gaudy wallpaper and shiny brass knobs and locks.

"When men leave to work in the mines," he continues, "they pay one shilling to be ferried across Lake Nyasa in a *dhow.* When they return months later with their earnings, the boat's owner charges ten shillings to cross the lake. Supply and demand."

We drive south all day, through thick bush and up and down hills; in the high country there are a few cows herded by African children because it's too cold here for tsetse flies. The road itself is extremely rough but not wet; previous mudholes, now dry, have logs and branches imbedded in them indicating where trapped travellers have struggled to extricate their vehicles—an omen of our future.

"Roads are a new phenomenon in Africa," Martin says. "Before Europeans came the wheel was unknown; everything had to be carried on people's backs or heads, so paths worked fine."

"I guess we're lucky to have the roads we do," I reply, rolling my eyes as we jolt along.

"Yes. Blessings on the British colonials," he laughs.

We stop for a lunch of tinned fruit and crackers by a ford, over which a rush of water about a foot deep is cascading.

"There's been lots of precipitation in the hills today," Martin comments mildly. He's used to travel in Africa. We wait for half an hour for the flood to subside, then ford the dam by driving across it through an inch or so of water.

At five o'clock we stop at a Rest House similar to the Dak Bungalow in Itigi, which costs us each eight shillings for room and the cooking of food.

Martin gives the African man in charge a tin of stew, which he heats up over a wood stove because there's no electricity. We eat in the large dining room lit by an oil lamp, then have baths in water labouriously heated in buckets by our waiter.

The next day we visit several more stores for Martin to check out the soap question.

"There are huge signs proclaiming Sunlight and Lifebuoy soaps, but they're the only kinds stocked," I say. "Why waste money on advertising?"

"Retail is in flux," Martin explains. "There used to be two large chains of stores called Mandala (eyeglasses) and Kondodo (walking stick), named for the distinctive characteristics of the two Europeans who founded them, but their trade is gradually being eclipsed by small Indian stores. Europeans impose a twenty-five per cent markup to ensure a decent standard of living for themselves, but Indians can make do by charging only ten per cent on goods."

On the porch of one store a tailor is sewing a man's suit on a treadle sewing machine.

"They'll tackle almost any order," Martin says. "I get all my shirts made to order because stores don't usually stock them. Shops that sell material often rent space to tailors for their machines."

"You can get a wedding dress made for five shillings and sixpence!" I exclaim, reading a listing of prices hung up behind the tailor. "That certainly beats costs in Canada." We're chatting right beside the tailor, but he doesn't look up from his work.

"And Tom-Toms (cheap cigarettes) are four for only a penny!"

"Pity we don't smoke," Martin laughs.

These tidy shops owned by Indians won't last. Eight years later, the first president of Malawi (Nyasaland's new name after its independence), Hastings Banda, berates the Indians for taking advantage of Africans. Later he denounces the Indians again; those who don't take the hint and leave have their stores burned down. The shop area here remains empty and derelict to this day.

"Africans aren't raised to be businessmen," an amused, stout Malawi bureaucrat tells Paul Theroux in 2001. "We have a much freer existence. Shops are not our strong point." He mocks the Indians as being obsessed with numbers and inventory and money, always one two three, one two three. Theroux is not amused.

Martin and I drive on south through mountains bordering the west side of Lake Nyasa (now Lake Malawi), the scenery marvellous except that now we can only glimpse it through a drizzle of rain, rain that will drastically affect our progress. Soon we're stopped by a five-ton lorry that has become stuck in a culvert and is blocking the now muddy road. The African driver and his assistant are futilely filling buckets with dry earth from the roadbank and spreading it, along with branches, under the front wheels. When the driver tries to move the vehicle forward, his wheels whir frantically, embedding themselves even deeper into ruts. We watch them struggle.

Soon a second, smaller lorry with a crew of four approaches from the south, and is also forced to stop because of the accident.

"Get a rope and fasten it to the lorry's axle," Martin orders the Africans from this second vehicle, who are now milling about on the road.

The six men immediately do this, using four thicknesses of rope to attach the two vehicles, Martin overseeing their efforts. The two vehicles both grind slowly forward, but when the rope is fully taut it snaps, the large lorry settling back into its old ruts. Its driver then decides to put the motor in reverse, but in doing so his vehicle retreats even further off the road, ending with its back wheels in a small stream at the bottom of the culvert. Fortunately this new catastrophe clears enough room for our car to pass and Martin drives on, circumnavigating mud holes with finesse as we go.

"Shouldn't we stay to help?" I ask.

"No point. They'll extricate the lorry sooner or later."

"How can you get away with ordering people around like that?" I ask. "You don't even know them."

"They expect it because I'm white," Martin answers. "If I was polite they'd think I was weak. If I didn't order them about nothing would get done; we'd still be back there sitting in the mud." The large lorry finally reaches Nkata Bay three days later, when we are starting south again after the weekend.

In Nkata Bay, Martin has acquaintances among the nine Europeans who live there; we'll stay for the weekend with Jerry Turner, a cheerful police inspector, and his wife and child. The wife seems unable to cope with Africa in that she and the child are never in evidence. Jerry gives us a tour of their house.

"You can have this room," he tells Martin and me, leading us to an airy room with a double bed overlooking the water.

"Oh," I say in a panic, looking anxiously at Martin with my hands clasped under my chin.

"We'll take two rooms, if that's okay," he says without missing a beat.

"All right," Jerry responds, startled, looking at Martin and then at me.

Being a good host for Jerry is difficult when there is no store nearby to buy food. Instead, supplies come every eight days on the boat *Illala*, which steams up and down Lake Nyasa, or visitors such as ourselves bring food with them (although we haven't done this). The Turners' two "houseboys" have tried to grow a vegetable garden, but with only limited success.

Martin wants to visit a friend nearby, so he drives off while Jerry and I go for a swim in Lake Nyasa, although I'm worried about safety. "There was a crocodile here a while back and later a hippo," Jerry laughs, "but they haven't been around for awhile."

"No bilharzia?" I ask before wading in.

"The water's too rough here for the snails that carry it," Jerry says.

"It's lovely and cool," I say, paddling about close to shore.

"We get long leaves here for health reasons," Jerry says, smoothing back his wet hair. "There's a lot of malaria around and biting tsetse flies, but at least they don't carry sleeping sickness."

On our stroll back to the house, we meet Martin, red-faced and annoyed, because his car has become stuck in the mud. We spend the next two hours working to free the vehicle. At first we pull up grass and lay it under the wheels, but in spinning they quickly reduce it to fragments. Martin gives the houseboys a cigarette each to cut more grass, but this doesn't work either. Finally Jerry borrows a chain and pulls the car to hard ground with his Jeep.

Before dinner we have drinks with six of the nine European inhabitants of Nkata Bay. Drinking is what they do when they aren't working, primed on this occasion by Jerry's liberal supply of liquor. Nick is a six-foot-six-inch Norwegian working on a construction crew; since he doesn't know much English, he's one of the more affable members present. Keith fancies himself in a moustache, beard, and waistcoat. He and John, both employed by the Public Works Department, sing a risqué song and then Martin and John's wife, who is also from South Africa, sing their country's national anthem. Jock, a middle-aged bachelor and by now drunk, pats my knee and calls me "honey" every few minutes. Then he massacres a poetic rendition of Robert Service's "The Shooting of Dan McGrew."

"When I first came to Nkata Bay, I used to explore the countryside around here and even paint some of it," Keith tells me after a while, as if suddenly sensing the feeling of malaise in the group, "but the climate gets to you. I don't have the energy any more. I just work, drink, and sleep."

"The biggest excitement around here is the arrival of the *Illala* every eight days," John says sarcastically.

"Then we can drink on water rather than on land," Jerry laughs.

"Remember the time you dove into the water in your underwear from the top deck?" John's wife says. They all laugh but Martin, who doesn't really like Jerry.

After dinner Jerry, Martin, and I stroll over to the African resthouse to watch a native dance, guided there by the steady throb of loud music. About two hundred Africans sit on benches against the walls, stand by the doorways, or peer through the windows. The women wear dresses with scarves over their heads, the men shirts and long pants or shorts. Some of the men have neat patches sewn in patterns on new shorts or trousers, giving them a personal touch.

At one end, by the dim light of three lanterns, is the three-man band on a platform, one man playing the accordion, the second shaking gravel in a cocoa tin as if it were a maraca, and the third pounding a steel hammer on a steel plate, his steady bangs largely drowning out the accordion's melodies. A fourth man collects two shillings six pence in total from would-be dancers— as few as four or as many as twenty—before the band begins playing each eight-minute set for them. Most dancers are solo men in thrall to the music— dipping and gliding, or bounding and leaping as the rhythm moves them, all proceeding in the same circular direction. A few single women walk around the room, several couples sway with the women hanging limply in their part-ners' arms, and rarely two men dance together. I've never seen people danc-ing by themselves, or two men dancing together, so I'm amazed. The dancers are serious about the activity—they've paid good money for their eight min-utes—yet carefree and uninhibited in their movements. I wish that Euro-peans danced this way, which of course millions of them will be doing by the end of the century.

A man brings three chairs so we can sit down against the wall, but we're largely ignored by those present. We're considered only intruding voyeurs, peo-ple staring at us when we arrive and later on as newcomers show up, so there's no thought that we might join the dance. The crowd is in constant flux as indi-viduals leave or push in after beer-sampling sessions outside; each adult is allowed to taste beer from the many batches brewed in the native quarters nearby before deciding which type to actually buy in a glass.

Sunday is a big rest day for the Africans, many of whom attend a church service at which Boy Scouts are singing. Most are dressed in their best, many strolling along the road under black English umbrellas or multicoloured beach parasols to keep off the sun. Some of the men walk in pairs holding hands, dis-cussing events of the day; men rarely walk with women, and never hand-in-hand.

Jerry, Martin, and I go snorkelling at Crocodile Beach but the water remains scary and the fish nondescript. We soon lie down in a row in our bathing suits to soak up some sun. Two women with bundles of firewood on their heads stop dead and drop their loads at the sight of us, chattering to each other and giggling. Why would anyone lie with so little on? And in the sun? Are we quite mad, if not dissolute? We finally gather up our things and drive the women up the hill on our way home.

After lunch, we set off to shoot hippos who are destroying African rice paddies at Chinteche, twenty-five miles away. After Jerry, now in his police uniform, "does" dinner by telling the house boys what to prepare for the meal, we drive in his Jeep to the police station to pick up another constable armed with a gun. The bumpy track leads past many huts, corn fields, and a rubber plantation, where people wave back when I wave to them. One house has two storeys, with the second floor reached by an outdoor portable ladder. At the time I note in my journal that the builder "forgot" to put in an indoor staircase, which is not only racist but ridiculous; an outdoor ladder is an ideal way to save room inside the building.

The river at Chinteche, about fifty yards wide, is running so high that the ferry isn't working. We have to cross the water in a dugout log of a kind still in use today, poled and paddled at each end by an African man. The two constables and Martin hunker down in the log, but my hips are too wide to do likewise; I have to hunch above them, feeling absurd. At the far side the police confer with a group of Africans. It seems that the hippos from Lake Nyasa start up the river, walking along the river bed, at about three p.m. each day and only reach the rice paddies much higher up well after dark. Jerry decides we can't wait this long so we return home by log and Jeep. We've had a pleasant outing at government expense.

The next morning I visit the Nkata Fisheries Research Lab where Dr. Derrick Iles, from Cardiff, is the first biologist to study the fish of Lake Nyasa.

"The lake is a goldmine," he tells me, excited to be talking to a fellow zoologist. "I've caught ninety different kinds of fish, mostly cichlids. I've found about six new species or subspecies that all live in the same place. Isn't that amazing?"

"How could they have evolved that way?" I ask.

"Exactly. But they all have different habits, and the natives call them different names when they catch them."

"They certainly look alike," I say, gazing at two silver fish with different name tags lying side by side on a counter.

"This one usually has seventeen soft rays in the dorsal fin—80 percent of the time anyway. And this one has only sixteen soft rays 85 percent of the time," Dr. Iles says proudly. These are distinct subspecies I ask myself?

"You have a lot of work ahead of you," I say, thinking of the number of fish he would have to catch to work out such percentages.

"I don't mind. I'm engaged to a zoologist so there'll soon be two of us."

Dr. Iles was indeed onto a goldmine. He became a world authority on cichlid fish, in 1972 co-authoring a definitive book on the group. Well over two hundred species of cichlid fish have by now been discovered in Lake

Jerry Turner and Martin English being poled across
a river in a dugout canoe, Nyasaland.

Malawi (Nyasa), due to its great age, large volume, and enormous depth (seven hundred metres); many of these cichlid fish have become immensely popular among owners of aquaria. After eight years of work on the fish in Lake Nyasa, Dr. Iles studied the population dynamics of Atlantic herring, first in Britain, and then for the Canadian government. He has been awarded various high honours for his discoveries, and although retired since 1997, he continues to do research from his base in New Brunswick.

I stroll over to the police station and listen to a discussion between Jerry and an African who contracts out jobs for the Public Works Department.

"I want to import eighty men from Tanganyika to cut trees," he's telling Jerry. "They work a lot harder than the men around here." He and Jerry consider the problems involved in bringing in labour when many local men are without work.

After lunch Martin and I drive to a nearby rubber plantation, where the Scottish manager, Mr. Shanks, shows us around. The rubber trees in the eight hundred acres are getting old, so Mr. Shanks will soon have to plant young ones. Each tree has a one-foot-square cut in the bark of its trunk, where every second day "sap boys" shave off layers of accumulated latex weighing about five ounces. Each man is responsible for four hundred trees. The latex is left with a little formic acid to solidify for a day in flat aluminum pans. Then it's put through rollers squeezing it flat, hung up, and packaged in bundles to be sent to factories that make crepe soles for shoes.

On the proceeds of this business Mr. Shanks and his wife live like royalty. When Martin and I drive to their place over a mile away for tea and drinks, we admire their large house situated at the top of a hill, with a magnificent view of the lake and surrounding rainforest. Below are a tennis court, a golf tee-off area, a fish pond, and a large garden where flowers, bananas, pineapples, tangerines, avocados, mangoes, and paw-paws flourish.

"I guess Scotland was never like this," I comment. Scottish farmers I've met are usually making do with small farms, stone fences, and ancient buildings.

"Mind your feet," Mr. Shanks says suddenly.

I glance down at my shoes, then up at him to see what's wrong.

"You're standing on an ant trail."

When I look more closely at the ground, I see thousands of tiny red ants streaming through the grass and beginning to climb up my trousers.

"Yipe," I bark, jumping sideways.

"They're the kind that bite," Mr. Shanks explains. "They climb all over an animal and then all bite at once. If the animal can't get away, it's had it." I quickly brush them off my legs.

"We don't have any children," Mrs. Shanks says, ignoring my panic, "but we have two dogs, two cats, and lots of ducks and hens. Sometimes a leopard visits at night, and there are lots of monkeys about."

Jerry has invited to dinner two animated White Fathers from the local mission, both dressed in white flowing cassocks with large rosaries around their necks. One is from Montreal and one from Boston. We chat about sports and Canada and snow while the "house boys" quietly bring out food from the kitchen and carry away empty plates. What a pleasant way to entertain!

When they leave, Father Shirson leaps into the sidecar cabin attached to their motorbike, then leaps out again to tell Jerry something at the last minute. His colleague drives off with a flourish and when we wave frantically to indicate that his passenger is missing, he thinks we are waving goodbye and waves back. It takes several minutes for him to realize that Father Shirson is indeed gone. He returns, collects his colleague, and then the two men tear off in a cloud of dust, their white robes streaming out behind them.

Aside from the British colonial staff itself, missionaries are the only group that provides social help to the people of this land-locked country, which is one of the poorest in the world. After independence, Malawi will gradually become flooded with foreign aid charities: Poverty Crusade, Action Aid, People to People, Save the Children, UNICEF, UNESCO, Mission against Ignorance and Poverty, Food Project. In 2001, Paul Theroux meets representatives of these agencies everywhere, being driven in pairs in white Land Rovers or white Toyota Land Cruisers by a black driver. All of them refuse to talk to him or give him a lift—how different from my trip when white befriended white at every opportunity so as not to undermine the colonial imperative. Theroux comments that, in spite of these many agencies working for many years and spending billions of dollars, some of them staffed by dedicated foreigners, living conditions for their "clients" have not improved and have often worsened. He sees many of the agencies as boondoggles that welcome large-scale disasters as "growth opportunities." Their fatal flaw is that they are run by foreigners using foreign equipment, with little advice accepted from local people.

Martin and I leave Nkata Bay the next day to continue our drive south over a high mountainous road. We haven't left the mire behind, though. When we come to a very muddy stretch and face to face with a bus going north, the bus driver promises to push us back to the road if we'll get out of his way. This we do. As we inch off the track, our car slides sideways as if on ice. It takes several men from the bus to help us regain our familiar ruts.

With no stores along the road and no food in the car for lunch, I become obsessed at noon with the thought of eating bananas. Martin isn't hungry, but eventually stops so I can buy some. Over Martin's objections I purchase twenty-four at half a penny each to make sure we won't starve. Famished, I wolf down one banana and then a second, but by the third I'm full.

"Ahem," says Martin glancing at the pile of fruit beside me and giving me an "I told you so" look. The bananas quickly rot in the hours and days ahead, filling the car with stench and me with guilt for my gluttony. Eventually I throw them away.

We stay the night at a rest house at Kasungu where three men are already ensconced, one a geologist and another an insurance agent; five people at once is something of a record at this place. There are only two bedrooms, so I have one while the four men have to share the other. We eat dinner in the third room, the men often talking animatedly at once; there are so few Europeans in Nyasaland that everyone knows everyone else or at least has acquaintances in common. And Africa and its inhabitants are perpetual topics of interest.

"I heard one native telling another that white men eat snakes," one man says out of the blue. "He'd seen a tin of salmon and thought it was snake meat."

"There used to be a man-eating lion here," says Martin.

"There was a rumour that Europeans had poisoned brown sugar," says another man. "For three months the Africans bought only white sugar, even though it was more expensive. The government had to issue a statement denying that there was anything wrong with the brown sugar."

"There are some stores in Salisbury where you can't buy sugar at all unless you speak Afrikaans," says a man.

"I know," says Martin. "I've heard that President Strydom in South Africa thinks one way to annex Southern Rhodesia is to have a slew of Afrikaners emigrate there."

The next day we drive eighty miles to Lilongwe, where Martin has to put in a day's work on soap-related matters. I relax in my room at the best hotel in town, which costs me forty-five shillings. When I put my suitcase on the bed and open it, a three-inch cockroach steps out, scuttles down the bedspread, and slips under the bed. I glimpse another, smaller cockroach under my packed shoes, but rather than pursue it I leave the suitcase lid open, hoping it will join its friend. I write letters sitting on the other end of the bed, and later stroll around town. Martin and I have dinner with two friends of his, who chatter away happily about personal affairs all evening while I pretend to listen, tired and bored.

Not far out of Lilongwe early the next morning, Martin and I come to an immense morass of ground-up mud extending two hundred yards; a bus and a lorry are stuck in the middle of it up to their axles. We get out of our car to see if we can inch past them, as a few other European drivers are doing.

"Last month we had twenty-two inches of rain near here in twenty-four hours," one man says. "No wonder it's still muddy!"

"I commute along here every day," another remarks. "Usually when I get to bad parts I have to pay natives ten shillings to push me through. It's a good way for them to make money, I suppose." There are about a hundred African men in ragged clothing lining either side of the road, waiting for just such an opportunity, but with luck we manage to slip past the lorry and bus without manual help. These men are also paid to clear away rocks by hand, keeping the storm drains open in the process.

It seems impossible, but when Paul Theroux drives over such a Malawi road forty-five years later, it's in even worse shape—more rocks, more mud, larger holes, deeper ruts. By 2000, donor bulldozers are being used to push aside rocks, which then often block the drains. This means that, when it rains, the rushing water isn't diverted from the surface of the road and instead washes it away. Such is progress?

Soon we reach the Liwonde River, where a two-car ferry attached to an iron cable takes us across it; the boat is propelled forward by six African men pulling on a rope as they pace from the front of the boat to the back. While I'm taking a picture of them, a toothless old African woman backs into me by mistake. We both laugh and grin at each other for the rest of the short trip, while Martin rolls his eyes at our lunacy.

Anne on the Liwonde River.

African crew pulling the ferry across the Liwonde River, Nyasaland.

As we continue our drive south over hilly roads, the number of vehicles and pedestrians increases. Martin notices one of his sales trucks approaching and stops it, berating the driver because, even though the temperature is over ninety degrees, he isn't wearing the regulation Levers Brothers overalls. Then he says a few words of Gujarati to the man, who is from India.

"I was asking after his wife," Martin explains to me when we set off again. "I'm doing quite well in the Gujarati lessons and in the Sales and Transport course I'm taking. I'm going to Europe in six months to take more courses in England. I get £120 a month and a company car, so I can save lots of money."

"You know exactly where you're heading," I say in awe. I wish I could say the same for myself. What will I do when I have to leave Africa and the giraffe, my life's ambition accomplished at age twenty-four? Maybe Ian will marry me and maybe he won't, but either way I know I'll want to continue working as a zoologist. Will I be able to land a job in Canada? There aren't many women scientists there.

"In two years I'm going to get married and we'll live in a garage I've already designed. I'm going to start up a transport business near Salisbury, where my wife will look after the chickens and run the flower shop."

"You've never mentioned a girlfriend!" I say with surprise.

"I haven't met her yet," he laughs. "I have two years to locate her."

We come to a Tsetse Fly Station, built to prevent vehicles from carrying tsetse flies into the fly-free region we're approaching. It's a big aluminum shed beside the road through which all vehicles and pedestrians must pass during the daytime; at night, tsetse flies don't follow moving objects so the shed isn't used. When we've driven inside, the doors at either end are shut so that the only light present comes from a side window. Any flies attached to our car are supposed to fly toward the light and be caught in a bottle by an African man. Another man cursorily sprays our car inside and out with DDT. We don't know at the time how dangerous DDT will prove to be in the future.

"We didn't go through a shed when we came into tsetse land farther north," I say to Martin, wondering how effective the system can be.

"That's because the road goes too high and it's too cold for the flies, forming a natural barrier," he explains.

We drive through Zomba, the capital, stopping for tea at the Zomba Hotel, whose verandah has a glorious view over plains rimmed with mountains. "Those are the Mlanje Mountains," Martin says, a range I'd read about in Laurens van der Post's emotional book *Venture to the Interior*. In 1949, van der Post had set out to survey them with thirty native bearers.

"Imagine exploring with porters on foot only eight years ago!" I comment. I don't know then that van der Post was infamous for embellishing his tales and making up stories, as described in the 2001 book, *Storyteller.*

We continue the thirty-five mile drive to Limbe over a tarmac road crowded with other vehicles and pedestrians. Blantyre, named after David Livingstone's birthplace, is only five miles from Limbe, home to both thousands of Europeans, including Martin, and many thousands of Africans, so we feel as if we're truly back in civilization.

Had we thought about it, I (but probably not Martin) would have assumed that, with independence, Nyasaland would be able to develop steadily into a thriving nation, unimpeded by colonial rule. Alas, this was not to be. In 2001 Paul Theroux revisited the school near Blantyre where he taught in 1964. He remembers it as a fine institution, the eager students in uniform, the library with ten thousand books. At present the school is still functioning, but barely. The buildings are battered, the trees cut down, the lawn grass chest high. Windows are smashed, walls covered in mildew, roofs broken. The library is almost in total darkness with all but one light fixture empty, but this doesn't matter because all the books have been stolen; millions of dollars in foreign aid money for education has also been embezzled. He believes the schools remain poor because of government policy; if they are destitute, they will continue to receive foreign aid, which can be siphoned off by government officials.

Before dinner, we go to the airport so I can buy a ticket for the two-hour flight to Salisbury the next day. Then I collect my mail at the post office, and send off a telegram to Mr. Matthew to tell him where I am and when I will arrive in Umtali, Southern Rhodesia, where he has proposed we meet. I've decided to fly to Salisbury despite the expense, because both train and bus service can take up to a week.

In the evening Martin takes me out for farewell drinks at the huge Limbe Club, crowded with sophisticated whites, and for dinner at the Flamingo Restaurant in Blantyre, where I'm booked into the Shire Highlands Hotel. Such fancy hotels that charge me a few dollars for the night have today upped their prices to $250 a room, which can readily be extracted from foreign non-government organizations. This is more than the average Malawin earns in a whole year! What a travesty this foreign aid money doesn't go as intended to African poor, but instead to rich whites for one night's accommodation. Giving charity money to groups helping Africans, obviously, is not necessarily the best way to go. In neighbouring Zimbabwe, which has been too prosperous until recently to attract NGO officials, hotel prices are far more reasonable.

In the morning, Martin drives me to the airport so I can board my plane at noon. In my journal I write that Martin "is one of the nicest people I have ever met, and exceedingly kind." And he never even made a pass at me! I correspond with him for a few years after I'm back in Canada, but we eventually lose touch so I don't know if his plans for a perfect life came true.

On the plane, an only partly full Dakota, I'm pleased to have a window seat since we're only flying at 180 miles an hour, and at an altitude of about six thousand feet. I realize how impossible it is to comprehend the vastness of this green bush country from a car, and equally impossible to appreciate from the air the might of the Zambezi River from its thread-like appearance. The two stewardesses pamper the passengers with a magnificent meal of two kinds of cold meat and a cold leg of chicken (which I wrap in a napkin to serve for my supper), before we land at the Salisbury airport, one of the busiest in Africa.

A free bus takes me to the city proper and a taxi to the train station, where I check my two bags. With seven hours to wait for the train going southeast to Umtali at the border of Mozambique, I wander about the city in a daze, marvelling at the myriad cars and the huge number of white people. No one says hello to anyone else, nor wonders why this is so. White women don't slop about in sandals like I do, but dress despite the heat in high heels and smartly tailored suits and dresses. African women wear shabby cotton dresses in contrast to those I saw in Tanganyika, who looked radiant wrapped in bright cotton cloth.

When it's dark and after I've collected Mr. Matthew's letter from the post office, which assures me he will meet me in Umtali, I settle down in the Ladies Waiting Room at the station. With me are a young, desperate Afrikaner couple and their five small children, some sitting on the floor amid their parcels tied with string. I haven't seen poor whites like this since I left South Africa. Two of the children lug the baby around to keep it quiet, putting it down outside on the platform where it crawls along in the dirt, checking out cigarette butts.

"Where's the Ladies lav?" a white woman asks me anxiously.

"Over there," I point.

"But there's a man in there," she wails.

I shake my head in sympathy and she sadly retreats.

An African woman staggers into the room with a suitcase balanced on her head and a boy of about ten strapped to her back, his feet dragging along the ground.

"He's had polio," the white station master explains to me when he sees me staring at them.

"Isn't that a rather primitive way to travel?" I comment, thinking that surely she should have access to a cart.

"Of course," he replies, "but the boy's legs are paralyzed so he can't walk; she's taking him to the hospital."

Later I linger on the platform waiting for the train to start after I've deposited my luggage in my sleeping compartment. Several white men run baggage carts along the platform, a job done by Africans further north. A young white man wanders over to talk to me.

"I'm a train fireman," he says. "It's a hard job. I have to keep my foot steady on the pressure gauge. One of my friends who fell asleep on the job was killed when he fell off the train."

"How awful," I agree, "but I thought Africans were firemen, not white men."

"Natives are no good," he says. "They couldn't do my job. If they could, then they'd be allowed to but they can't. Not smart enough."

I feel like telling him that Tanganyika has native firemen, but don't. What's the use?

"I'm a passenger on this train," he says. "Do you want to come to my compartment to talk when we get going? I have to stay awake until midnight when I get off."

"No," I say, "I'm tired," which is true.

In my sleeping compartment, I chat for a few minutes when we're in bed to my young, white companion, who is going to Umtali to visit her boyfriend.

"I have to work in Salisbury because the pay's too poor everywhere else," she explains. "My father has a farm fifty miles outside Salisbury and he also runs some trading stores; it's hard for him since he has to hire wogs and *kaffirs* to work for him. They're bone lazy and rude. Except for the old, respectful *kaffirs,* of course. They're okay."

I roll my eyes in the darkness and soon fall asleep to the rattle of the train car, not waking up at midnight when we stop so the conceited fireman can get off.

# 16

## Zimbabwe and Victoria Falls

When the train pulls into Umtali at 7:30 a.m., Mr. Matthew is waiting beside his ranch wagon to meet me. We greet each other with concealed delight, not even touching to shake hands, then have breakfast at a hotel where I tell him about my adventures. He has been touring orange orchards in Southern Rhodesia in connection with his consulting business.

Umtali, an attractive and popular resort centre, is located at the extreme east of Southern Rhodesia. From there we drive south through the eastern highlands into a drier region, the number of baobab trees increasing as the number of pedestrians fall. Around the Sabi River, there's evidence of an extensive irrigation scheme set up for Africans, but we see only a few squatting by the road selling basketwork and handmade mats.

As we drive, Mr. Matthew tells me about what has been going on in South Africa. "The missionaries who service rural areas are up in arms because of apartheid," he says. "At each outstation they used to have a room reserved for them in the house of some important African, where they could sleep when they came to visit their flock. Under apartheid, it's illegal for a white person to sleep under the same roof as a non-European, so now they have to build huts or sleep in their cars.

"There's also a fuss about the five hundred Hungarians who were allowed to immigrate to South Africa last year after the Hungarian Revolution. South Africans were very proud of themselves for being so generous. Now, citrus farmers like me are being pressured to hire as many of them as possible because most don't have jobs, despite all the rhetoric about giving them a new chance in a new country."

At Fort Victoria we follow a dirt road south toward Zimbabwe, the remnants of a once-famous habitation (now known as the Great Zimbabwe National Monument) which gives the country its present name. We book two rooms at Sheppard's Hotel, which advertises itself as "Near To and Facing" the Zimbabwe Ruins, a precise description although they are two and a half miles away. We then drive to the site, encompassed in well-kept green lawns with an adjacent nine-hole golf course.

The magnificent stone structures of Zimbabwe were built before the European Middle Ages by skilled Bantu stonemasons, probably of the Shona tribe. Most of the remains, situated part way up a long, high hill, comprise a compound, the Great Enclosure, with outer walls over thirty feet high that narrow upwards because no mortar was used in their construction—the stones were chosen and shaped so as to fit exactly together in neat rows. Where the original walls remain intact, there is a fancy chevron pattern in stone near the top. Most striking within the walls, sheltered by a large shade tree, is a solid conical "temple" made of stones, whose purpose is unknown. Some authors suggest it is a phallic symbol. There are also lower, inner walls forming narrow passages through which we squeeze.

Below this central compound are extensive piles of rocks extending far off into the distance—apparently the remains of dwellings. Far above, at the top of the hill, is the Acropolis—a fortress built with massive boulders ingeniously connected with stonework. Tiny passages and openings between the boulders, among which we clamber, have been given names such as the Balcony and the Gold Furnace—gold hunters in the 1890s stole gold from here, wrecking many of the structures in the process. From the highest boulders I can see a vast distance over the surrounding bushland, where hills and granite outcroppings poke up through the green vegetation.

Zimbabweans are justifiably proud of the Great Zimbabwe showing, as it does, that they have an impressive inheritance. Archeologists conclude that from about 1100 AD to 1500 AD the site was an important religious and secular centre with 10,000 to 20,000 inhabitants and a vast communication and trade network extending as far as Botswana, Mozambique, and South Africa. The city was abandoned before the coming of the Portuguese, probably because of a loss of grazing possibilities, of soil fertility, and of timber in the surrounding area. The Great Enclosure through which we wander may have been built by a king for his mother and wives. Contrarily, because of the great height of the Enclosure's outer walls, archeologist Wilfrid Mallows believes that the most likely possibility is that it was a prison built before 1500 by

Anne in a passageway of the Great Enclosure of the now-called Great Zimbabwe National Monument.

Anne at the Acropolis of the Zimbabwe Ruins.

Bantu natives (probably Shona), where captured Africans of other tribes could be collected before being shipped off into slavery in the Middle East.

We stay two nights at the hotel so that we can give Zimbabwe the attention it deserves. However, I'm sidetracked for a few hours by a chameleon which we pick up from the middle of the road after screeching to a stop to avoid hitting it. It's over a foot long with a long curved-under tail, a ridged back, and stalked eyes that rotate autonomously in a startling manner. On the ground it was brown, but when I hold it near a green bush its skin turns green, like a large leaf complete with reddish veins. I keep it in my room overnight, hoping that it will snag the flies there with its extensible tongue. In the morning, however, it's in a foul mood, turning black and hissing fiercely. What a scary world it would be if our human moods were similarly apparent! I immediately set it free.

The hotel is lovely, with open patios made of stone and large rooms with thatched roofs. The other guests are mainly Afrikaners, including thirty schoolchildren.

"It was the Arabs who made Zimbabwe," an Afrikaner man insists, against all evidence. "There's no way *kaffirs* could have made something that good."

One woman who has been drinking at the bar with her husband and two children falls flat on her face beside me in the lobby. She moans "I feel awful," several times, and does not even try to walk as her husband and I drag her to their room. Her children gather up her purse, cigarettes, and shawl without being told to do so by their father. I'm much more upset by this mishap than are her family members.

Mr. Matthew has the brilliant idea that we should take a detour before heading back to Fleur de Lys to visit Victoria Falls, even though it's well over four hundred miles away in the opposite direction.

"There are giraffe near there," I remind him.

"Oh yes, giraffe," he says, although I'm not sure that he knew this. "Those will be a different race I think?"

"Maybe," I say, although I don't really know this. "Certainly new giraffe terrain. I'd *love* to see them."

"Let's do it then," he agrees. "We'll drive to Bulawayo for a late lunch, then book a train seat for the three-hundred-mile trip to the falls. It's too far to drive, and there's nothing much to see on the way."

"It would take too long by car," I agree.

"The car might break down, which would be bad in the middle of nowhere."

We drive to Bulawayo on strip roads—asphalt or cement bands a yard wide, running parallel a few yards apart so that vehicles' wheels are always riding on these strips unless a car or lorry approaches from the other way. Then both vehicles move left so their left wheels shift onto dirt, and they can pass each other.

Bulawayo is nearly as large as Salisbury, with equally wide streets reflecting Cecil Rhodes's admonition that a full span of sixteen oxen should be able to turn around on them. Pretty bungalows set in flower gardens abut the many parks. But the past and future history of this attractive city belies its air of innocence. Beginning early in the last century, in Bulawayo as elsewhere, Europeans stole productive farmland from the Africans, forcing Bantu men into hard labour in mines or factories or plantations. Resistance by the Bantu against the tyranny of whites increased over the years, especially after 1965 when the white government of Southern Rhodesia would declare unilateral independence from Britain so that the Bantus could continue to be denied the vote. For fifteen years the area would serve as a battleground, with thousands of blacks slaughtered because they demanded freedom. More recently still, the black Zimbabwe government will make war on its own people. Young girls will be snatched off the streets of Bulawayo and sent to paramilitary camps, where they will be raped by young men conscripted by the government to stifle political dissent among Zimbabweans.

After lunch we drive to the train station to buy tickets but find, to our chagrin, that the train to Victoria Falls is sold out because it's school holiday time.

"Oh dear," I say. "That's too bad when we've come so far."

We drive to the Automobile Association for information and maps. The one elevator in the building is reserved for Europeans, so Mr. Matthew rides in it while I walk up the stairs in a small boycott gesture. In the AA office we are heartened to learn that there is a strip highway all the way to the falls, and that road conditions are good.

We set off at once, driving through dense bush until darkness finds us at Halfway Hotel in the middle of nowhere, along with, surprisingly, about fifty other guests. I'm given a room complete with five large spiders lurking in the wainscotting so, despite the heat, I avail myself of the mosquito netting over the bed for anti-spider protection.

The next morning we arrive at Wankie, a large coal mining town with neat houses and gardens, where we buy supplies for our visit to the falls. As we're traipsing about the streets I notice that the post office has separate outdoor entrances for natives and for Europeans, so I wonder where the Indians go. In the butcher shop there are two queues, one for three Europeans served by two men, and the other for about fifteen Africans with a single man waiting on them.

As we are crossing over a railway bridge we come upon two African men walking along together, one wheeling a bicycle. When he notices us the man without the bike, startled, steps backward off the bridge and disappears. His friend walks on with his bike, looking surprised and shrugging his shoulders at the funny white people when we stop the car and back up to see if we can help the fallen man. When we look down at him in consternation, twenty feet below us, he's standing rubbing his ankle. He looks up at us and grins, surprised that we should have been worried about him.

The falls announce themselves in the distance with a column of mist rising several hundred feet into the air. Then we hear the roar of rushing water as we cross the bridge below the falls into Northern Rhodesia, becoming soaked with cooling spray in the process. Magnificent.

Mr. Matthew is highly organized; although we plan to stay at the north bank rest camp, we drive immediately to the nearest hotel to make reservations for a game flight for £5 each. There's a convention of Rotarians there, some of whom know Mr. Matthew and come up to him to chat. They try to hide their shock that he is here with a young woman from Canada rather than with his wife. Then we drive back to the tea room on the north bank to book seats on a launch trip upstream.

These tasks completed, we settle down at an outside table to sip cokes. We gape at the mile-wide river rushing past in front of us before dropping out of sight to our left, the widest expanse of falling water in the world, and at the many small islands and plants clinging for dear life at the top of the falls as the water swirls by at their roots. The water level is as high as has ever been recorded, twelve feet above normal, so the scenic drives and benches along the river are completely underwater. A brazen sign, barely rising from the water surging past like an express train, reads "Canoes for hire." Another says "Crocodiles, bathing is suicidal," but who would swim, either crocodile or person, where the likelihood of being swept over the falls is certain?

We drive to the south bank and walk down a trail parallel to the edge of the falls into a rainforest, deafened by the roar and soaked to the skin with spray. This must be the epitome of rainforests because, although the vegetation isn't incredibly dense, the "rain" never stops. At openings through the trees and bushes we take pictures of the face of falling water and of the rainbows emanating from it, hoping that the spray won't ruin either our photos or our cameras. Victoria Falls is 350 feet high, twice as high as Niagara Falls, so we aren't nearly far enough away to film the complete wall of water.

On the way back to our temporary home, we visit a giant baobab called, naturally, The Big Tree, which is enclosed with a fence and guarded by an African. He sits beside his charge all day to ensure that none of the few passing tourists dares to carve initials in its trunk.

Once back at the rest camp, where we each have a square, thatched hut with bars on the windows, Mr. Matthew hires an African lad to look after our meal which he will share, albeit at a distance, with us. The boy tells me he can't cook, so I heat up our canned stew. I ask him to bring some hot water for coffee, but he says he isn't allowed to do that, so he sits and watches me lug a bucket full of water from the women's washhouse. He does do the dishes, but in the process loses the big bar of soap Mr. Matthew has given him. How can he lose a bar of soap in a bucket? He makes me grind my teeth in annoyance.

We're up before dawn the next morning so we can photograph sunrise behind Livingstone's statue at the south edge of the falls, here called the Devil's Cataract. David Livingstone was the first European to see this wonder (my guide book says that he "discovered" it in 1855), which he named for his Queen. Mr. Matthew practises walking backward with his movie camera so he can film the statue, more and more of it becoming visible as he retreats, with the sunrise radiating out from behind it.

After breakfast we clamber down the side of the river below the noisy falls, this time on the north side, again through dense rainforest blocking out the sun. Mr. Matthew is behind me so I soon feel, romantically, that I'm alone in a primeval rainforest present since the beginning of time. At least I do until I come to two men leisurely sweeping the steep path free of leaves with twig brooms. At the water's edge I crouch down near piles of boulders to put my hand in the brown water swirling around in a large eddy. Near a palm grove, a horizontal liana hangs beside the path at waist height. I take a moving picture of Mr. Matthew, sweating profusely in the sauna-like heat, leaping onto the liana and swinging back and forth. Given his weight, it's a very strong vine.

Once back at the top of the trail, soaking wet from sweat and spray and exhausted from the climb, we drive east along the Gorge Road to glimpse the river as it threads through a series of switchback gorges. We buy a brick of ice cream and cold cokes for lunch, the only tourists in sight.

"In this heat it's easy to see why Africans haven't got a complex culture," Mr. Matthew gasps.

"True," I pant. "But then there's Zimbabwe!"

On the Internet, fifty years later, I read that entrepreneurs have initiated bungee jumping off the 360-foot-high Victoria Falls bridge and white-water rafting through these gorges; well-heeled tourists pay large amounts of money for the thrill of charging through twenty-three separate sets of winding rapids.

In the afternoon we drive three miles west up Livingstone Road to the dock, where a flat-bottomed boat with a roof and open sides, and room for

Statue of David Livingstone at Victoria Falls.

about thirty people, is tied. We're early, so Mr. Matthew chats with the white driver, who says he's a keen photographer with a licence to shoot two thousand crocodiles in the Okovango Swamp in South West Africa. He used to live near Klaserie, so they exchange gossip about common acquaintances there.

"I used to farm near Klaserie, too," a woman with three children joins in, overhearing their conversation and moving eagerly up the boat toward them.

Immediately the men fall silent, with nothing further to say to each other or to her. The woman, rebuffed, returns to her family.

"I don't like talking to strange women," Mr. Matthew explains to me later when I ask him about his odd behaviour.

Before we cast off, a few more tourists with cameras and two men from Durban join the boat, the latter strutting about in shorts to show off their tan and their muscles. We motor slowly upstream near the shore for about two miles, the river as still as glass. The high water level is a drawback—there are few animals in view because of ample drinking holes inland, and the Game Park we pass is closed, its fence half submerged in water. I'm disappointed at this because I've read in the tourist literature that a giraffe called George lives here, who eats buns out of your hand. It's too hot even for crocodiles to lie on the sand banks, although we notice three small ones back under bushes. (A woman asks of no one in particular, "Are crocodiles man-eaters or fish-eaters?" but no one answers her.) We do see baboons, a drinking kudu, cormorants, and fish eagles, but no hippos.

The following day we drive to the airport for our game flight. Eight of us squeeze into a biplane, the heaviest at the front. I'm pleased to be deemed the lightest and therefore the last to climb aboard. But if I'm the lightest, and not really that light, surely the plane will be too heavy?

I remain nervous as we swoop over the falls a number of times, banking first to the left and then to the right so we can all have a good view. Sometimes, to my horror, the pilot stands on a wing, pointing out highlights below.

"Don't forget to lean with the plane when it banks," he announces when he's back in his seat, the last thing I want to do. As we head back toward the airport I worry that something is wrong, we're going to land, but instead we soar just over it, barely missing the trees beyond. In an open patch, we come in a great roar upon an elephant, who looks up at us aghast before hastening off in fear. I know how panicky it feels.

"We were only ten feet above the ground!" I exclaim to Mr. Matthew later. "I saw individual leaves on the bushes!"

"Nonsense," he replies. "I could see the altimeter and we never went below three hundred feet," which still seems to me impossible to believe.

With our plane's shadow flitting along below us, we now home in on whole herds of animals—not only giraffe but also roan antelope, zebra, eland, ostrich, waterbuck, warthogs, and thousands of sable. Mr. Matthew had told me that the sable at Fleur de Lys were special because there were few left in the world, but this obviously wasn't true. I look at Mr. Matthew, but he's twisting in his seat, absorbed in trying to film antelope through his small window. Some of the animals flee from us in terror as the elephant did, but others glance up curiously to see what is making the racket up there. I find myself looking across at the herds rather than below, which gives the impression of travelling in a bouncing car, albeit at a speed of a hundred miles per hour. I snap photographs in my enthusiasm, but none turns out well.

After about twenty minutes of this excitement I begin to feel queasy, a sensation not helped by the woman in front of me who sits holding her head and moaning, rather than looking out the window. I clutch a barf bag in my hand, praying I won't have to use it.

When the pilot swings the plane about and announces "We're heading back, now," I'm greatly relieved until Mr. Matthew points out a group of elephants he's missed to the pilot, and to my horror we dip, roll, and bank over these animals too to get a good view.

"Enough, please enough," I mutter urgently under my breath, clutching my bag at the ready.

Fortunately, we then fly straight to the Zambezi River and follow it eastward, seemingly ten feet above the water, for the fifty-mile trip home. After we land we stagger out of the plane one by one, stiff, nauseous, and elated. In two and a half hours we've flown over Northern Rhodesia (now Zambia), Southern Rhodesia (now Zimbabwe), Bechuanaland (now Botswana), and the Caprivi Strip of South West Africa (now Namibia) and seen not only giraffe but many thousands of other animals in their native habitat. Who could ask for anything more?

Even so, it's still light so I explore the riverbank to try to spot a hippo, one of the few common species we haven't seen. I fight my way upstream through thick reeds and bush, transected by muddy trails covered in hippo footprints, which lead to the water. Completely constricted by the vegetation around me, I listen nervously for any sound of animal life but hear nothing.

"Apparently buffalo are seen along there as well as hippos," Mr. Matthew says lightly when I return. "And if you frighten hippos on land, they escape by rushing along their trails to the water."

"Knocking over anyone in their way, I'm sure," I comment.

"Actually, a native was killed recently by a hippo while collecting firewood on one of the islands."

"Thanks a lot for telling me now," I laugh sarcastically.

Over a late dinner Mr. Matthew and I worry about the race of the giraffe we've seen from the plane.

"There's one called *Giraffa camelopardalis thornicrofti* in Northern Rhodesia," I say, looking at notes I made back at Rhodes University, "but it's at the other side of the country. Near here they could either be the *angolensis* race, which is to the north, in the Barotse area of Northern Rhodesia, or the *capensis* race to the west, in Bechuanaland."

"What's the difference? What do they look like?" Mr. Matthew asks.

I refer again to my notes. "*Angolensis* has spots that are slightly notched and go right down to the hooves, and *capensis* has spots that go down to the hooves and which are fairly round, but sometimes with pointed extensions."

We look at each other and laugh. "Surely that's six of one and half a dozen of the other?" Mr. Matthew says.

"And actually, there's no boundary such as a mountain range or a wide river between the two, so they probably interbreed anyway," I continue.

"How would you describe the spots of the giraffe we saw from the plane?" Mr. Matthew teases me.

"Spot-like?" I laugh, thinking of the animals flashing past my window.

"They looked like the giraffe at Fleur de Lys to me."

"Those are supposed to belong to the *wardi* race," I say, "but I agree. I couldn't identify any race of giraffe with certainty from only its spots, except for *reticulata*."

"What about vegetation? Bechuanaland is much drier than Fleur de Lys," Mr. Matthew comments. "Is that significant?"

"Probably not. Giraffe used to have a much wider range than they have now, what with European settlement and farming, so we can't really tell if they like dry areas or if they only live there because they've been pushed out of greener regions." Even today there's uncertainty about the races of giraffe in this region of Africa. In southern Africa the *wardi* and *capensis* races are now grouped together generally as *Giraffa camelopardalis giraffa,* using the subspecific name given by Europeans exploring northward to giraffe they first sighted after crossing the Orange River.

The trip back to Fleur de Lys is steady driving, beginning at 3:30 a.m. when we pack the car and set off from the rest camp. Mr. Matthew says he never sleeps more than four hours at night, so I don't feel guilty about falling into unconsciousness as soon as we wheel onto the strip road. We breakfast with other white travellers at Halfway Hotel where we had stayed on the drive north, Mr. Matthew commenting on races of people rather than giraffe.

"There's a touch of the tarbrush in that group over there," he comments, nodding at a family of four. The father is darker than his wife, and their gorgeous young children have very curly hair.

"But not enough to get them evicted from the dining room?" I ask, still amazed that one's skin colour should matter so much.

"A lot of Afrikaners have African blood," he says. "It makes it awkward for the powers that be who decide who will have second-class status. If families have lived long enough as whites, there's probably no problem. I doubt that they spend much time in the sun, though."

From Bulawayo south the highway, now called the Great North Road, is fully tarred, which makes easier driving. However, this is cattle country so now and then we have to slow down to cross a cattle grid—parallel, round steel rods embedded at either side of the road at fairly close intervals so that vehicles can cross over them but cattle, with their narrow hooves, cannot. The grids function as fences without having to be opened and closed. We pass a Frey Bentos factory, the name familiar from corned beef hash tins sold in Canada; the company owns one of the huge cattle ranches of the region.

The only town of any size on our route is West Nicholson, a mining centre. The railway line from the north stops here because of historical disagreement and a difference in rail gauge; it was originally intended to go eighty miles further to Beitbridge, so that Rhodesian and South African goods could be readily shipped back and forth. Both countries have lost business because of this gap in rail service, and now Zimbabwean goods are often shipped by train to the ocean through Mozambique.

We spend the night at Beitbridge, the customs town beside the Limpopo River. While Mr. Matthew is arranging for our rooms, a man with a moustache asks me for drinks with him and his friend. When I refuse, he becomes abusive.

"Your dad won't mind," he insists.

"He's not my dad. He's a friend," I snap.

"He's far too old for you."

"He's just a friend. I don't want a drink."

"I'll find your room and come and get you tonight," he persists with a leer.

"Then I'll call Mr. Matthew," I snarl.

As Mr. Matthew and I drink a lemonade before dinner, the man sits on the verandah behind Mr. Matthew, raising his glass in a salute whenever I glance his way. He makes me very nervous. What if he does break into my room? Will Mr. Matthew hear me call out? I want to tell Mr. Matthew what the man has been saying, but that would bring into the open the matter of sex, which I'd rather not do. That night I lie awake for an hour, listening for any unusual noise. Mr. Matthew is in the next room, but as I can hear him snoring I still feel vulnerable. Nothing happens all night, so I have missed sleep for nothing.

Our final day on the road begins with customs to cross the brown Limpopo River—still not green or greasy, despite Kipling—into South Africa, where the terrain again becomes hilly and there are many more people in evidence. Around Pietersburg, Ndebele Africans wearing thick copper anklets on each leg trudge at the side of the road. Some of the women sport full native costume while others are clad in ragged dresses. They don't seem to mind when I take their pictures. After a long day's drive we arrive at Fleur de Lys in the late afternoon, to be greeted by a large giraffe standing in welcome by a waterhole.

I've covered an immense distance in my travels and seen different races of giraffe and their habitat, information that will be useful in my future work describing the distribution and subspecies of this animal throughout Africa. Even so, I'm thrilled to be back where I can recognize giraffe by name, and watch Star and Pom-Pom and Cream going about their daily lives.

# 17

## Back at Fleur de Lys

Now that I'm back at Fleur de Lys, I immediately begin again to observe and record what giraffe are up to. My main focus, however, becomes completing the giraffe film, and adding more sequences to the 16 mm colour footage that Mr. Matthew and I have already taken to make it into a complete documentary of the animal's behaviour. We have visions of showing it publicly in a bid to conserve land for wildlife in Africa, or of selling it to Walt Disney to make money to help wildlife. I already have good sequences of Star walking, Cream galloping, an extended close-up of Cream trying to wrap his long black tongue around a high leaf while feeding, sparring between Star and Lumpy and several other males, giraffe drinking at a waterhole and at a cattle trough, and Star chewing and swallowing his cud. Mr. Matthew, during my absence, has filmed splendid slow-motion shots of single and herds of giraffe running, and a scene of giraffe silhouetted against the setting sun.

New giraffe calves have been born at Fleur de Lys since I've been away, so I'm thrilled to be able to film a mother suckling her young. In fact, I'm so excited that my hands shake as I try to change the reel to take even more footage and I manage to jam it in the camera; I have to wait until I'm back at the farmhouse at night to fix the problem in my blacked-out bedroom. On the way home, annoyed at my stupidity with the camera, I drive carelessly over a sand ridge on the track, onto which Camelo becomes stuck. I have to walk four miles to the Fleur de Lys office to ask Mr. Matthew if he will allow some workers to drive back to Camelo and lift her clear. He agrees, but I feel badly to be so inept.

It's important to show that the development of land for farming in the eastern Transvaal has not destroyed the lifestyle of the giraffe, so I spend a day trying to film in slow motion a giraffe stepping over a four-foot wire fence—a sequence that will later prove invaluable for a scientific paper I'll publish called "The role of the neck in the movements of the giraffe"—because they swing their necks far back when lifting their forelegs and well forward again to lift up their hind legs and finish a "jump." With the film rushing through the camera at four times its normal speed, it's distressing to have my target animal suddenly pause reflectively beside the fence instead of crossing it, perhaps even mesmerized by the noise of the camera. I finally have my shot, but at the expense of reams of useless footage.

"When the farm was first fenced years ago," Mr. Matthew tells me, "the giraffe routinely walked through the fences and broke them. The damage was so great that the board of Fleur de Lys talked about shooting all of them. But then they somehow learned to go over instead of through the fences, and most of the damage suddenly stopped."

An African is still employed to do nothing but mend the endless miles of fencing on the farm, but lapses on the part of the giraffe are rare. One giraffe became so fence-conscious that he walked up to a wire gate that was open without noticing that there was no wire there. He lifted first one foreleg and then the other high into the air, edged forward, then pulled his hind legs up and over after them. He walked calmly on, ignoring the shout of laughter behind him.

In the middle of the day when the light is too harsh for photography, especially when it's very hot, I edit the acceptable short clips into our definitive film using Mr. Matthew's film editor and glue. On the evenings when processed films and slides are returned to us by mail, Mr. Matthew, George, and I screen the former to see if they're good enough for the master reel. For newly developed slides, we set up the slide projector to examine and admire them. Once, to my horror, I realize that I've taken a photo of a male baboon sitting on a rock with his pink penis exposed, the full monty. I rush on to the next slide, worried that one of the men will say something, but they don't. I learn later that this posture in monkeys and apes has a name, a penile display, meant to intimidate other males and maintain dominance. I often think of it today when I see sitting men holding court with their legs spread apart, a posture uncommon for women.

Our cinematic group also discusses what other characteristics of giraffe need to be documented in the master film. We're all eager for slow-motion shots of giraffe lying down and getting up, which I finally obtain after much wasted footage—it's as difficult foreseeing when a giraffe will change its position as it is when one will jump a fence.

Giraffe at Fleur de Lys.

We decide that the film should have some human interest—maybe something about me coming to Fleur de Lys to study the giraffe. Mr. Matthew films Camelo bounding up and down hills along the main road and through bushveld areas of the paddocks. He has me climb out of the car, go up to a tree at which a giraffe has been browsing, and tie a ribbon around it.

On the evenings when Mr. Matthew is away, I work on the cattle report Mr. Matthew has asked me to prepare, organizing the extensive notes I've made on what plants I've seen them browse, and how they divide their time between browsing and grazing. I also catch up with the newspapers he has saved for me in a large heap. I'm keeping a scrapbook of political clippings, reading each new development in apartheid with amazement:

- Now there's a new law that fines anyone who allows white nurses to accept orders from non-European nurses C$200, as if this is likely to ever happen, given the racial structure of the country.
- The Minister of Native Affairs announces that any native can be refused entrance to a church in a European area, which is most of the country and where one-third of Africans live. What sort of Christianity is this? Dutch Reformed Church, apparently.

- The English-speaking whites are worried about the Prime Minister's refusal to attend a meeting of the Commonwealth Council of Prime Ministers, as well they might be, seeing that the country will soon vote to become a republic.

Mr. Matthew has told me to my relief that since it is now April, the beginning of cold weather, snakes will be hibernating. This turns out not to be true, which is hardly surprising when the temperature still reaches 100 degrees Fahrenheit at midday. One Sunday, when Mr. Matthew and I are on foot hunting for giraffe to film, a cobra flushed by a moving herd of cattle comes through the grass toward us, with its head in the air and its yellow hood spread. Frozen in place with horror, I stand watching it approach, unable to move or even call out. Luckily, Mr. Matthew sees it and rescues me by grabbing a thick stick and quickly killing it.

A few days later, when I'm driving slowly near the shopkeeper's house in Klaserie, a four-foot cobra streaks across the main road right under Camelo, spreading its hood before diving into the bush. For once, safely ensconced in my car, I'm not afraid of the creature.

Nobody on the farm much cares about snakes, but the same is not true of lions. While I was away a young giraffe was attacked and partly eaten by lion on Fleur de Lys property, and now a pride of eight has moved onto the adjacent farm and killed nineteen cattle, a loss of over $1,000 for the farmer. Since the lions hang out each day in dense bush where they can only be hunted on foot, all the men are afraid to go after them.

"Two men up the road spent a whole afternoon hunting lions, their rifles cocked at the ready. They got separated by accident, and one lost his way and had to sit up in a tree all night. His family was more scared than he was, and he was pretty scared," Mr. Matthew tells me.

Now and then lion tracks are sighted on Fleur de Lys property, and one night lions kill two heifers and a sessaby. Mr. Matthew arranges for the fifteen hundred head of cattle to be corralled late each afternoon in *kraals,* an enormous task, so they can be guarded all night by African men sitting up over small fires.

One evening, when Mr. Matthew is away and I'm invited to dinner at the home of two tennis-playing brothers and their families whom I met six months ago, I tell the group about our lion troubles. Bob and John grow tomatoes on their farm as a cash crop, but also have wild animals in their unfarmed bush, including elephants, and sometimes lion and cheetah. One African, two weeks earlier, had speared a leopard with his *assegai.*

"There's a pride of lions in the area right now," Bob says. "One farmer down the road lost six cattle within a week to them."

We're having coffee after dessert when John suggests we drive around the farm by way of evening entertainment to see if we can spot some lions in the car headlights.

"They'll probably see us coming and clear out, but I'll take my gun anyway."

"Can you just shoot a lion from a car if you see one?" I ask. This seems highly unsporting.

"Of course," John says. "Lion are vermin. The fewer there are, the better."

Six of us and John's toddler, held by her mother, Iris, crowd into their car and drive slowly along the tracks that transect and circle their property. We stop for a few minutes near a waterhole where nothing is happening, and are on our way home when a male lion crossing the track ahead of us stops to stare at our lights.

"A lion," John hisses. "Get the gun. On the floor in the back. And get a torch."

By the time we wrestle the gun to the front seat through our bodies and find two torches in the car that work, the lion has moved calmly on into the bush. When we drive around the bush area we find him again about ten yards away, completely unafraid although lit up by two torches and headlights from our car.

"Gimme the gun," Bob whispers, leaning over John's shoulder. "I've got a better shot."

"No. I can get him from here," John hisses back.

"Why is daddy shooting the pussy?" the toddler asks in a loud voice.

"Hush up," snaps her mother.

"Gimme some room," grunts John, pushing Bob away from him and taking aim at the lion.

At the loud bang, the lion grunts and leaps forward out of sight.

"Damn, you missed," cries Bob.

"No, I got him. Look, there's blood on the ground."

"We can still find him," Iris says.

"Should we get out and search?" I ask.

"No way. Not for a wounded lion," Bob retorts. "Not in the dark."

We drive around aimlessly for a time, peering into the darkness with our torches and the headlights, but there's no sign of the animal.

Back at the house the men arrange to meet in the morning to follow up the hunt.

"We have to get him. It's too dangerous to leave a wounded lion at large. He could kill somebody."

"Can we get the Game Warden to come? He has the best gun," Bob says.

"He isn't allowed to shoot vermin, only game animals," John answers. "That's a bitch 'cause he's paid from our taxes."

"I'll come too," I say. I feel sorry for the lion, but with the rush of adrenalin from the evening there's no way I want to miss the excitement of the next day.

I drive to the farm in Camelo at first light the next morning. John has his truck parked by the house into which we all pile—in the cab are John driving with Iris, their child, and John's mother, and in the back Bob and two other white men with guns who are obviously able to take time off work at the drop of a hat, three African men with an axe and an assegai among them, a young African *ayah* to help with the toddler, five mongrels owned by the African men, and me. When we arrive at the scene of the previous night's shooting, the women and child settle down around the truck while the men and dogs set off into the bush, following the blood trail of the lion.

The two white women and I are chatting idly when suddenly there's a terrific noise of growling about a hundred yards away in the bush. We dash for the cabin of the truck, diving in headlong with the African *ayah* on top of us. We roll up the windows and sit shaking, wishing the men and guns were here to protect us. John's mother pulls a small revolver out of her pocket.

"I'll shoot any lion in the eyes," she announces bravely, waving the gun around.

Soon it's quiet and so stifling hot in the cab that we roll down the windows and then slowly emerge to sit around the truck again. Fifteen minutes later we hear four shots and the roaring of a lion from about a mile away. Fifteen minutes after that there are three more shots, more growling, and hysterical barking from the dogs. Then three more shots. What on earth is happening? How many shots does it take to kill a lion? After all, it was already wounded.

After a while John bursts out of the bush to our right.

"Surely the lion's dead by now?" calls Iris. "We heard a lot of shots."

John gives her a cold look. "We have him cornered in a dry creek," he says. "We need the truck to go after him. You all wait here."

"No way," says Iris. "We're going too. I'm don't want to see a lion shot, but I'm not going to sit out here with lions all round."

"At first we disturbed five lions feeding on a wildebeest just over there," John explains as we clamber into the truck, Iris, his mother, and the child in the cab with John and the African girl and me in the open back.

"You must have heard them growling," he says as he starts the motor and we slowly jostle forward. "We found the wounded male further on. He charged us so we all shot, and one bullet hit him. Then he bounded off. He can't have been hurt much last night. Then we tracked him, shooting when we could, and now he's at bay. The natives are sitting in a big tree on the other side of the creek to see if he makes a break for it. The dogs are huddled under the tree; the lion nearly got one of them, so they won't have anything more to do with the hunt."

John guides the truck in first gear through the bush, bumping over stones and logs, knocking over small bushes, and detouring around trees as the *ayah* and I, crouching in the back, cling to the sides of the truck to avoid being flung out. When we reach the other men sitting in a tight clump, they climb into the back so that now we have four guns (John has his in the cabin with him), the mother-in-law's pistol, and my camera, all at the ready.

John edges along parallel to one side of the creek and then back along the other side, all of us silent and tense in case the lion suddenly leaps out at us. Could he jump right into the back of the truck? Would I remember to take a photo if he did? Should I then leap out of the truck? Will any of the men help me or are we all on our own? We peer into the bush but it's too dense to see much. When John crosses the creek bed for the second time, the truck becomes stuck in the sand. There's a long silence.

"Anyone want to get out and push?" John asks. No one does.

John guns the motor but the wheels dig deeper into the ground.

"We'll have to go in from here," says Bob. The men with guns slowly approach the creek in a line, shooting into the bush at intervals just in case the lion is there. The rest of us watch nervously from the truck. At last they see the animal lying part way up the creek bank and finish him off. Immediately the tense mood of the group changes to jubilation.

"He's huge, and in fine condition," exults one of the men.

"His pelt's pretty good," says Bob.

The African men first push the truck out of the sand, then lug the dead animal to the vehicle and sling him into the back, forcing those of us already there to step aside. The dogs whimper and push against my legs as we jounce toward the house, upset at the strong lion smell.

At the farm house the carcass is offloaded under a large tree where it's measured—nine feet from nose to tail tip—and will later be skinned by the Africans. All the men take turns having their photograph taken with the lion. One tracker puts his leg in the animal's open jaw and sprawls on the ground with his other leg in the air, pretending he's dead. The African men were the bravest during the hunt because they crossed the creek without guns, only twenty yards from the lion, to reach the tree on the other side, but no one comments on this.

"The boys scrape off the lion fat into small bottles to sell to witch doctors for half a crown each," John remarks as a group of us stands looking down at the animal.

"Lions have a small bone in their thigh muscle that isn't found in other animals," Bob says to me. "Would you like to have it as a souvenir?"

"I'd be thrilled," I say, although I'm puzzled. I've taken several courses in anatomy but never come across anything like a floating leg bone. The cat doesn't have such a structure; why would it have evolved in the lion? And why only one when the lion has two thighs? However, all the men look pleased at this suggestion and my response, some even laughing with delight.

"I'll get the boys to cut it out and you can have it after we've had tea at the house," Bob says solemnly. As I'm leaving in Camelo an hour later, he presents me with a bone about three inches long. When I get home, I put it outside on my window sill where ants make quick work of the bits of flesh attached to it. I save it in a metal tin, only realizing years later that it is in fact the penis bone of the lion, the *baculum,* and that the men were playing a joke on me.

A few nights later, a lion leaps into one of the Fleur de Lys *kraals*, grabs a calf in her jaws, jumps back over the brush fence with her prize, and bounds into the bush. The men watching over the cattle can only stare in amazement because farm workers aren't allowed to shoot guns.

"She probably had cubs to feed," Mr. Matthew says philosophically.

The cattleman, Mr. Van Vechmar, decides to poison the lions to get rid of them. He shoots a wildebeest and drags the carcass behind his Jeep around the bush area of the farm so that any resident lions will smell the bait. Then he salts the flesh with arsenic capsules, and waits for the predators to drop dead. Unfortunately, a neighbour did this same thing with a calf carcass some days earlier and although the lions ate some of the flesh, the arsenic made them vomit, so none died as far as anyone knew.

I hear about the lack of success the next day when I'm talking to Farnie and Van Vechmar in the office. "They came to the bait. They even dragged it a few yards, but they didn't eat any of it," Van Vechmar says with annoyance.

"What will you do with a carcass full of arsenic?" I ask helpfully. I haven't seen the wildebeest but can imagine it rotting in the sun, bringing death to any scavenger who samples it.

"I'll set a trap for them," Van Vechmar decides, ignoring my question. "The lions probably noticed a human smell on the carcass that reminded them of the arsenic. I'll kill a cow under a tree and sit in the tree all night to shoot the lions when they come to eat it. Will you come too, Farnie?"

African farm workers who helped in the lion hunt.

"I suppose so," says Farnie after a pause; he likes the idea of getting out of the office, but isn't sure about losing a night's sleep.

"I'll come," I offer, not wanting to see a lion shot but anxious to taste the adventure; they ignore me.

In the morning I find that this plan was no more successful than the last. The two man, guns at the ready, were ensconced in the tree shortly after dark but no lion ever appeared.

"Farnie kept falling asleep and snoring," Van Vechmar says in disgust. "He nearly fell off his branch."

Farnie shrugs his shoulders. "Two jackals came and had a great feed as we watched," he says. "We couldn't scare them off because that would have warned the lions."

Maybe the lions were deterred anyway by Van Vechmar's efforts, because they no longer attack Fleur de Lys animals and soon leave the area. For several nights I wonder if we'll see a lion when Mr. Matthew and I drive slowly around the farm without our lights on, hoping to catch poachers who are shooting Fleur de Lys game and trucking it out to the railroad to be sold elsewhere as meat, but we encounter neither kind of predator.

At Easter, a citrus man and his family come to Fleur de Lys for the holiday. His wife lived for a time in Ottawa, where her father served as a South African representative to Canada, so we talk about Canada as well as South Africa. One day we drive to a forty-five-thousand-acre game farm abutting Kruger National Park, whose northern area is currently closed to the public;

it acts as a buffer against poaching in the Park. We encounter lots of animals but only a few species—giraffe, kudu, wildebeest, zebra, and elephant (represented by a heap of steaming feces). Mr. Whittingstall from Acornhoek, one of the three directors of Fleur de Lys, comes with us in his own truck. He is a pioneer of the lowveld, now over seventy years old.

"I've shot over a hundred lion in my day," he says offhandedly, "and only the last one got me."

"What happened?" I ask.

"Mauled some ribs. Not too bad." He won't say anything more about this attack.

The next evening, before the family leaves, we drive around Fleur de Lys property and see five giraffe whom I can't identify in the dusk, the sable herd, wildebeest, four sessaby, two duikers, six kudu cows, four young waterbuck, two secretary birds striding along looking for snakes to eat, and a multitude of cattle. We may not have as many animals as the game farm, but we have a greater variety which we all enjoy.

When I think of soon leaving this paradise in May, I'm filled with emotion. I become introspective about the past year. When I lose the nails of my two big toes, I think back to my thirty-mile trek down Kilimanjaro when, at each step, my toes jammed forward into my borrowed boots; over the next weeks they turned blue and black and finally are loose enough to fall off, revealing pink pristine nails growing beneath them.

My friend Julie's letters from Dar es Salaam make me laugh about our time together working for the government in Tanganyika. She writes that one day she went home, leaving the keys to the department safe on her desk by mistake. The Labour Commissioner saw them and, because she had become a Bad Security Risk, demoted her from being his personal secretary to the common typing pool. She's philosophical, noting that at least the Commissioner doesn't know about the many times we forgot to lock the safe at all. She says she and Paddy are planning to get married—I'm the first person to know.

The return of my raincoat from my cabin on the boat on Lake Victoria reminds me of how kind people were during my travels. A man I scarcely knew had gone to the trouble of packing up the coat, taking it to the local post office in Kisumu and mailing it back to me at his own expense.

I'm looking forward, too. Griff writes that she likes my report, but that I should separate what was known before about giraffe and the new things I've discovered. I had thought that I should include everything that was known, which wasn't much, but Griff explains that this isn't the way science works. Rather, each researcher presents only his or her new data in each scientific paper to add to the previous body of knowledge; later on someone will synthesize all that is known on a topic into a review paper or monograph. I can easily rewrite my paper as she suggests on the two-week boat trip back to London. I dream that I'll be the one to write the definitive book on giraffe as well, which turns out to be the case.

My mother is planning to fly to London to meet me when I return there, so we can tour together around England and Scotland. I'll enjoy this immensely, because I share her interests. Before that, Mr. Matthew has decided that when I leave in Camelo I should visit Kruger National Park one final time, and then drive to Hluhluwe Park in Natal so that I can see the "tame" giraffe, Shorty: he's the reason I was lucky enough to come to Fleur de Lys. Mr. Matthew will follow me in his car, then carry on with business of his own when I finally set off for Grahamstown.

And Ian writes he is thinking of possibly coming to England in late August; if he does so, we can get married there so we can wander a bit through Europe together before returning home to Canada by ship; it is inconceivable that an unmarried, middle-class couple could sleep together in the same hotel room. Being married seems to me increasingly like a good idea—I don't know what I want to do when I leave Africa, I'm running out of money, and since my mother now lives in two rooms at the University of Toronto, I have no family home to return to in Canada.

A week before my departure, Mr. Matthew gives me a Zulu shield made from the hide of the young giraffe killed by lion, together with two Zulu *assegais* or spears, and a fly whisk of hairs made from the tail of the giraffe I dissected. All of these keepsakes were fashioned by Offis whose name, Offis Mathebula, is printed on the back of the shield, along with a salute to "Inkosikas-ka-Uhlu," which translates into "Laughing Lady," apparently my nickname among the workers. Finally he gives me a beautiful, brown wool blanket adorned with tawny giraffe and other animals, which continues to keep me warm, both physically and spiritually, fifty years later. He has Offis build a wooden crate in which to ship these presents to Canada.

"How can I ever thank you?" I say to Mr. Matthew, tears in my eyes because of his kindness.

"We were glad to have you," he says, embarrassed at my show of emotion.

Finally it's time for me to go. The eight-hundred-foot film is complete—not perfect, since with more time I could probably replace much of the footage with even better shots, but good enough to make us proud. Mr. Matthew invites the Europeans on the farm to a party to see it. During the day of the gathering I bake cupcakes in Watch's oven, and produce two batches of ice cream in an ice-cream-making machine, stuck at home because Mr. Matthew has arranged for Camelo's tires to be retreaded for the trip back to Graham-stown. At the party I meet wives for the first time, who watch politely with their husbands not only the giraffe film but also some of my slides and other movies that Mr. Matthew has taken. The next day I visit all my friends to say a final sorrowful goodbye, then pack up Camelo for the trip south to Rhodes University.

# 18

## *Leaving the Giraffe*

By driving east from Fleur de Lys, Mr. Matthew and I again visit Kruger National Park, this time with two cars; we'll drive south through the park rather than along the main road. After parking Camelo at the Satara Rest-camp, I walk over to join Mr. Matthew beside his ranch wagon. On the way I meet Kirk, our African ranger/guide of October.

"Hello," he says, glad to see me again. "You back. Where is the other lady?"

"Mrs. Cook couldn't come on this trip," I explain. "How have you been?"

"Good," he says, smiling. He goes to Mr. Matthew to greet him, too, but Mr. Matthew only grunts in reply.

"Don't you remember Kirk?" I ask Mr. Matthew after Kirk has retreated.

"Of course," he replies. "He's being smarmy so we'll give him a big tip."

"No," I retort, dismayed at this interpretation. "He's being friendly!"

"I'll never be the friend of an African," Mr. Matthew states.

I look at him in amazement. "Maybe Kirk's lonely," I say. "He probably wants to practise his English. Maybe he has no one to talk to in English?"

"I don't care if he's lonely. I don't care anything about him," Mr. Matthew states flatly.

We set off to view animals, I subdued at first because of Mr. Matthew's absurd behaviour, but then soothed by the glorious assortment of wildlife—giraffe, elephant, zebra, impala, buffalo, kudu, nyala, waterbuck, baboons, and the nose of a submerged hippo. We return to camp just before closing time at 5:45. A woman drives up to the rest camp several minutes later.

"It's not my fault for being late," she calls to the white man who has just locked the gate.

"Sorry, you'll have to pay the late fine," he says, turning back to open the gate for her to enter.

"But I was charged by an elephant!" she maintains.

"Oh," he says uncertainly, pausing as he unlocks the padlock.

"I would have been on time but I had to stop so he wouldn't attack my car. I had to wait for him to cool down."

"All right," the man concedes. "I won't charge you."

She drives past us with the importance of a person who has been charged by an elephant.

The next day, we head off in two cars to the Lower Sabie Camp fifty miles south. On the way, a man who flags down Camelo from his car jabbers at me in Afrikaans. When I shrug my shoulders in abashed incomprehension, he laughs.

"There are two lions by the dam ahead," he says.

He stops Mr. Matthew too, who satisfactorily answers "Ja."

Three cars are lined up by the dam, staring at two brown specks a quarter of a mile away. The man in the next car explains to me in Afrikaans where to look; I try to imagine what he's saying and peer there with my binoculars, but he soon switches to English. I loan him the binoculars to thank him for his help. Another man who has built padded armrests on both of his front windows peers at the lionesses with a camera lens a foot long, his wife relegated to the back seat to be out of the way.

The two lions are lying in the dust, snoozing. After a few minutes one lumbers to her feet, stretches, moves behind a bush and back, then flops again onto the ground.

"It's ridiculous," I laugh at last. "Five cars watching two such lazy creatures so far away!"

Mr. Matthew laughs too, but we keep on watching.

We depart the next day for Hluhluwe, Mr. Matthew driving off at first light to finish business at the Swaziland Irrigation Scheme and leaving the movie camera with me since I have no need to hurry. He tells me later that, cameraless, he met first an elephant and then a lioness, each sitting in the middle of the road, and finally a herd of about three hundred buffalo that block his way.

I dawdle along, caught in a long, dusty line of cars with inhabitants peering out the windows, seeing only common animals far away—nothing worth filming. When a large male baboon leaps onto Camelo's bonnet and calmly sits down, giving a bored yawn that reveals large white canines, he's too close to film. As I restart the motor he gracefully jumps down, sauntering into the bush on the other side of the road.

When I finally leave the Park I have trouble finding the route to Swaziland. My first choice of a road leads me to the customs post for Lourenço Marques in Mozambique. There, an officer directs me back to the correct dirt road leading south for fifty miles through flat bush country, dotted with a few native huts, cows, and goats.

Entrance to Swaziland is by a gate leading directly into the huge Swaziland Irrigation Scheme (nicknamed SWIG), which runs for miles south through bushland. Six huge bulldozers parade in a noisy circuit, picking up earth from an enormous hole near the road and dropping it further along for road improvement.

Mr. Matthew told me to put a piece of sticky tape on the main SWIG sign so that when he had finished his business he would know I'd passed by, but try as I may, I can't find any signs at all for SWIG. Finally I wave down a bulldozer and shout at the driver over the roar of his machine, asking for directions to SWIG headquarters. There I see many signs but since I don't have much sticky tape, I find Mr Matthew's ranch wagon and tie a yellow ribbon around its nose, thus marking it as a giraffe fodder plant.

The road south to Bremersdorp, where we will spend the night, is hilly and winding with a dreadful surface—sections not pocked with rocks and corrugations are covered feet deep in sand. When I hit such a sand patch at forty miles an hour, I lose all control of Camelo, exactly as if she's on a sheet of ice, and pray that she and I won't swerve off the road or skid over a cliff before she slows down to a stop.

I've been on the road a long time, so now I pray also that I won't run out of petrol (I've already used up the petrol in the can I carry with me), or water for the radiator. I look at rainwater collected in a pool beside the road. Would muddy water in the radiator be better or worse than no water? What will I do if I run out of petrol entirely? There are lots of children walking home from school beside the road, but will they be able to help me? Or do they even speak English? Thankfully, before Camelo breaks down, I arrive at Balagane, headquarters of the huge farm south of SWIG, which owns the only petrol pump in this eighty mile stretch of "highway."

Balagane is owned by a huge European man named Forster, who arrives by Jeep with seven young African boys as I'm buying petrol.

"Good to see you," he greets me heartily. "Having a good trip? Do you fellows want some cookies from the store?" he asks the boys, rubbing the head fondly of the lad nearest him. They jump about in excitement, obviously crazy about Mr. Forster.

"I used to run a boys' school in Mbabane," he says. "Several of these lads went there—great kids, all of them."

He hands some coins to one of them. "This one's a brilliant gate opener," he teases. The lad ducks his head and grins with pleasure, then races off with the others.

I'm astounded by the wonderful rapport between this man and the boys. I've never seen anything like it in South Africa, but of course I'm now in Swaziland, not South Africa.

The rest of the way to Bremersdorp is very hilly, which provides, once Camelo has struggled up each steep incline in first gear, superb views in every direction of pasturelands dotted with huts and small *shambas*. There are lots of Swazis on the road, some of whom wave back to me while some don't. There are also squashed corpses of bloated puff adders. Just before dark, I collect water from a small stream for the radiator, then drive the rest of the way to the George Hotel in the dark. Mr. Matthew arrives in Bremersdorp an hour and a half after me, having driven the same distance in half the time.

Next morning Mr. Matthew and I buy food supplies for the next few days and more petrol. His ranch wagon has a petrol leak, but he fixes this up with soap and a piece of sticky bandage that I supply. I dump radiator cement into Camelo's radiator to seal small holes there. Then we set off in convoy first to Gollel, along roads as bad as the afternoon before, and then back into South Africa where they're somewhat better. I choose the road for Piet Retief to the west by mistake, so Mr. Matthew comes roaring after Camelo to tell me I should be on the road south to Natal.

Natal looks more like the large farms I'm used to, growing sugar cane, pineapples, sisal, and eucalyptus trees on large plantations. Zulus stream along the roads, the men dressed in ragged tops and pants—no more sharp Swazi kilts—the women often in dresses with wide sashes. Some have their hair mixed with clay, and worn in coiffures built up at the back of their heads for a foot or more.

When we reach Zululand, a large area within Natal, pasture largely takes over again. So many Zulus are crowded onto what is called their "homeland" that the earth is stripped bare of trees, even on the tops of the many hills. In the valleys are the ubiquitous small huts and *shambas* with maize and grazing cattle.

Hluhluwe Game Reserve, its headquarters and rest camp situated on a hill up which Camelo has laboured, is a far cry from Kruger National Park. It's relatively small, about forty thousand acres, and fenced with wire and sisal plants through which animals sometimes escape. Those that do are quickly shot by neighbouring Zulu or European farmers. Because the reserve has no large carnivores, visitors are permitted, under the direction of native guides who join each vehicle, to leave their cars and approach animals on foot.

The sleeping quarters comprise a few large cottages and about fifty neat *rondavels,* each furnished with two beds, a washstand, a table, and cupboards. A Zulu man cooks the food Mr. Matthew has brought, a Zulu waiter serves it, and a third Zulu sweeps out the *rondavels* and acts as a houseboy. Whoever thought that housework could be so labour intensive? And be without women's participation?

The highlight of the park for me is my introduction to the giraffe Shorty, the reason I was able to arrange to work at Fleur de Lys. On our first foray in Mr. Matthew's station wagon out of the rest camp, just before dusk, we come upon him right beside the dirt road, nibbling leaves on a tree. Far from moving off at our slow approach, he comes over to the car, so close that I can only see his legs out of Mr. Matthew's window. Suddenly his head appears too, his long tongue rasping at Mr. Matthew's windbreaker. Mr. Matthew, annoyed at this sloshing, tries to push Shorty's head away but I eagerly lean over and give him my hand to anoint instead. Shorty licks it for about a minute, covering it with slobber and filling me with euphoria. Then he wanders off. Because he was captured when only a few days old and imprinted on people, he ignores the other few giraffe introduced into the park and prefers to spend his time near the rest camp where he can hobnob when possible with human beings, the species to which he feels he belongs.

The next day with Bon Course, our silent Zulu guide, we drive slowly along dirt tracks through varied vegetation—grassy slopes where zebra, wildebeest, and warthogs congregate and forested valleys with occasional waterholes catering to kudu, buffalo, and nyala. When we sit on a hill with a vast sweep of landscape below us to eat our sandwiches for lunch, we count forty-five zebra, thirty wildebeest, twelve warthogs, twenty-four impala, three waterbuck cows, a number of baboons, and seven rhinoceros. Mr. Matthew is ecstatic about the rhinos.

"They're white rhinos!" he exclaims, peering through his binoculars. "They're almost extinct! It's amazing to see so many together!" Bon Course gives them a cursory glance, then continues chewing on his ham sandwich.

"Let's go closer," I suggest. "We're nearly finished lunch."

"No, they're coming this way. We'll be able to watch them without scaring them."

We wait for half an hour in the heat, then an hour as the rhinos drift closer, Bon Course taking the opportunity to lie down on the grass on his back and snooze, his hat over his face. Unfortunately, when they are two hundred yards away, Mr. Matthew notes to his consternation that they have pointed upper lips, not square ones (their name is from the Dutch word for "wide," not the English "white").

Mr Matthew and ranger Bon Course at the Hluhluwe Game Reserve, Natal.

"They're not white rhinos after all," he says in annoyance. "They're common black rhinos."

We drive off and stop at the large Hluhluwe River flanked by a dense growth of lianas, palms, and other tropical plants, but don't see any crocodiles.

"The warden thinks a crocodile killed a giraffe near here. They found giraffe bones in the river bed," Mr. Matthew says. "It could have grabbed the giraffe's nose as it leant down to drink."

In mid-afternoon, when Bon Course spots two bona fide white rhinos in a distant valley, we park the car and walk a mile or so toward them. The animals are huge, six feet tall at the shoulder, so they have no need to be nervous at our approach, but they dither about anyway, acting like silly children. They dash here and there in little spurts, but not away. Once they run toward us, not actually charging, but when Bon Course shouts they stop in a panic.

Farther along, beside the main road, we see two black rhinos down a hill, smaller and darker than the white ones. We edge slowly toward them too, along with American tourists from an African Safaris Volkswagen van.

"I'm scared," a woman dressed in bright, orange-striped pants repeats with each cautious step forward she takes.

We don't go as close to them as we did to the white rhinos, because black rhinos are much more likely to charge. The guides protect us not with guns, which they aren't allowed to carry, but with pebbles they collect to throw at dangerous individuals.

On the way back to the rest camp, we meet the Safari Car again. This time it is parked near two female giraffe with their half-grown sons, the vehicle shaking with the inner excitement, cameras sticking out of every window although it's nearly dark.

"Rather small for giraffe," I comment condescendingly to Mr. Matthew as we navigate around the van.

"Decidedly," he laughs.

The next day, a European man approaches Mr. Matthew at the rest camp as we're about to leave.

"Could you please drive me to my friend's car, which has broken down?" he asks. "We set off earlier and I've left this person sitting by the road, so I need to get back quickly!"

Mr. Matthew agrees, somewhat curious by the impersonal use of "friend" and "person." When we reach the car, the "person" turns out to be a female aviator from Durban with whom the man is travelling around South Africa. Neither of them knows anything about cars, nor does Bon Course seem to, so Mr. Matthew takes over and finds a broken connection of some sort, which he soon fixes.

The aviator meanwhile is endlessly cheerful. "Isn't he clever," she exclaims about Mr. Matthew. "Aren't we lucky that he was at the rest camp?" She gives a little skip. "Aren't the birds around here lovely?" We look into the sky but don't see anything flying.

The man says to me out of the blue, "I've been to Malindi up in Kenya, and slept in the same bed as Grace Kelly did when she was making the movie *Mogambo.*"

On the way back from buying petrol at Hluhluwe Station, we drive over a cobra which slithers suddenly onto the road from the grassy embankment. Mr. Matthew tries to back up and kill it so it won't suffer, but without success. We clamber out of the car on the side away from the snake.

"Get a stick," Mr. Matthew orders Bon Course and me. He finds a thick one for himself, picks it up, and advances on the wounded snake, now lying in the middle of the road. It dodges several of his blows, hurt though it is, and shoots venom at its attacker, hoping to blind him. Several streams of the poison arc toward Mr. Matthew's eyes but, because he's wearing glasses, the poison merely slides down the lenses and drips onto his shirt.

I'm shocked into action. "I'm coming," I call. "I'll get a stick. Mind your eyes!" Although usually not a fussy person, my lack of glasses as a protection against venom makes me very particular about a choice of weapon. One branch is too thin, one too knobbly, one too short. Bon Course, too, is unable to find exactly the right weapon. When I look over to see Mr. Mat-

thew standing relaxed and know that the snake is dead, I grab the short
stick anyway.

"Where is it?" I demand, rushing forward.

"Too late," Mr. Matthew says drily.

"Oh," I say in mock surprise.

"I'm sorry I had to kill it; the poor thing was scared."

Again I experience in quick succession the extreme rush of adrenalin
occasioned by fear of a snake and the pathos of a dead one. This is the sixth
snake I've encountered in Africa, four of whom were immediately finished off,
a sad reflection of our inability to live peacefully with nature. The death toll
reminds me of my grandmother who, on the Ontario farm where she lived,
killed every snake she came across despite their importance in keeping down
the number of rats and mice in the barn.

"They have little legs that they walk on, and they're bad," she would say,
influenced perhaps by Adam and Eve's nemesis.

We three sit near a wallow to watch animals come to drink the next morning.
About fifty warthogs cavort about, some grunting their own importance, others sparring and squealing, making a terrific racket. Several approach us to get
a better look, then snort an alarm, at which all of them immediately trot into
the bush, only to return a few minutes later. Once Mr. Matthew mimics this
sound so well that they all again rush off. Eventually a black rhino comes to
drink but, when Mr. Matthew turns on his movie camera, it looks about in
puzzlement at the sound, then wheels quickly out of sight behind bushes. One
joy of waterhole-watching is anticipation—who will come next to drink?

In the afternoon, together with three other tourists, we stalk five white
rhinoceros on a hill. The male, standing by himself with ears drooping, allows
us to walk within twenty yards of him and take dozens of photographs,
although he shudders at each click of a camera. Bon Course has stones in his
hand to throw should the rhino charge, but it's obvious the animal's feelings
would be hurt by such rudeness. But what if we were lions? Was he by tem-
perament made for extinction? Is he so clever that he knows by our demeanour
that we'll be sorry for him and leave him alone? (Strangely, forty-five years on
and after much poaching of rhinos to obtain their horns to sell as aphrodisi-
acs, black rhinos are now much rarer than white rhinos, with only a few thousand left in the whole world.) On the drive back to the rest camp, we see
hundreds of nyala coming out in the dusk to graze.

After dinner Mr. Matthew gives each of the Zulu men tips of two shillings
and six pence. They accept the money gravely, but as the waiter leaves, he
shouts "Whoopee" and races away across the lawn.

Tourists photographing a wild white rhino at Hluhluwe Game Reserve.

On Saturday Mr. Matthew and I part with sorrow, he to go about his business and return to the giraffe at Fleur de Lys, and I to drive the one thousand miles to Grahamstown.

"How can I thank you enough for everything, Mr. Matthew?" I say with a huge lump in my throat.

"You can call me Oom Alex from now on," he replies.

After shaking hands I drive off sadly, the dull sky and bumpy road matching my sombre mood. Eventually the road to Durban becomes paved, so I sail along faster. Bush gives way to pastureland, and then to huge sugar cane fields where tiny train tracks cross the road to be used in carrying the harvested cane for processing. I begin to see Indians walking along beside the road as well as Zulus.

I return the waves of two young white men in a Morris who pass and fall behind Camelo several times as we speed along. As I enter Durban on a four-lane highway they pull up beside me, peering over at me and grinning. When they pull ahead and gesture for me to stop behind them, I decide our friendship has gone far enough. I give a final wave and drive around them; they pause, then turn sharply onto a cross street and accelerate away.

I drive along the south coast road from one fancy resort to the next until it is nearly dark, and I stop at Port Shepstone for the night. I give the African man who carries my suitcase into the Milton Hotel a shilling because I don't have any smaller coins. I think of asking him if he has sixpence change, then realize how ridiculous this would be. What am I coming to that I should even think of something so crass? The manageress's son, somewhat drunk, asks me if I'd like to go dancing after dinner in the nearby town of Margate but I decline, saying I'm too tired.

At dinner I sit at a table next to that of an ebullient blond woman, who is thrilled that I'm a zoologist who has travelled outside South Africa.

"I'm a bird watcher! You must have seen lots of birds!" she enthuses in an Afrikaans accent.

"Quite a few," I say casually, trying to appear modest rather than a liar.

"There are hundreds right here in South Africa, of course!"

"Of course," I agree.

As I get ready for bed I go over the birds I remember seeing—albatross, an ostrich, tick birds, a stork. A pitiful list. Laughable, really. Why have I noticed so few birds when there are hundreds around? It's good to focus on one topic, giraffe, to learn as much about it as possible, but surely I should be more aware of the bigger picture too, the complete environment of the giraffe?

I don't fully realize it at the time, but up until now I've felt myself to be infallible—as good as anyone at what I do. Now it seems I have a one-track mind! I'm dead tired from driving for so many hours, but my mind searches for more birds I've seen…flamingoes…secretary birds…before I finally drop off.

Years later, my deficiency is recalled when I'm quizzed by two students who had done graduate work on birds in different parts of Africa.

"During your giraffe studies, did you see fish eagles?"

"Hmmmmm. No, I don't remember them."

"Bee eaters?"

"No, I don't think so."

"Hoopoes? Lilac-breasted rollers?"

"I can't recall any of them," I admit. How embarrassing for a zoologist.

Later I also realize that I not only often have a narrow mindset in zoology, but one which jumps to conclusions. When I am working on a book of local Ontario mammals, Mary Gartshore, a student deftly drawing the different species to illustrate their appearance, remarks that black squirrels aren't necessarily black.

"Some have reddish tails or even white streaks," she says. I nod as if I know this, but am amazed. I've seen many thousands of black squirrels before, but obviously never really looked at them. I open my eyes and over the next months do indeed spot individual squirrels with patches of fur of different colours. How could I not have noticed this before? Why did I assume that because the squirrel was called black, it was necessarily always black?—particularly as black squirrels are only a different colour phase and belong to the same species (and often the same litter) as gray squirrels? It's obvious I'll never be an artist.

I recall a painting of a unicorn I hung up in my room when I was young, although the animal had a bloody wound in its side and was closely surrounded by a fence.

"Do you really want this picture on your wall?" my mother asked with surprise, focusing on the unicorn's desperate situation.

"Yes," I insisted, seeing only the beautiful animal rather than its injury or its captivity. Only when I'm much older does the context of the painting suddenly strike me—the lovely animal in pain and imprisoned—and I realize the full implications of the picture. I immediately take it down and hide it away. At least my deficiency is sometimes of use, as I can exist happily with household dust and clutter, which saves much cleaning time.

The next day I drive inland along winding, hilly roads through the tiny towns of Harding, Kokstad, and Umtata. It's easy to identify the African reserves—scattered, round mud huts with thatched roofs, sheep, cattle, and not a tree in sight. Indians and Africans often draped in blankets ride slow horses or walk beside the road; little boys herding goats hold out their hands for money and yell at Camelo as I pass.

These people in the Transkei look desperately poor, but their lives will get worse. Under the government scheme of Grand Apartheid, whereby 85 percent of the population of South Africa is to be legally squeezed onto 13 percent of the land area, this arid wasteland will become the "Jewel of the Bantustans," a "homeland" to which natives, in this case Xhosas, will be restricted. The Transkei will gain first self-government and then independence under an African leader who, in exchange for supporting government policy, will become personally wealthy far beyond the dreams of his constituents.

Despite this setback for the natives, the eastern Cape, where blacks and whites have been in contact for hundreds of years, will continue to function as a centre of black resistance. The mission schools have been training students to speak and write English well enough to enable them fight back against white domination: Nelson Mandela, his early wife Winnie, and Steve Biko, leader of the Black Consciousness Movement, all grew up in the eastern Cape. But resistance calls forth repression, and many Xhosa men will be assassinated by the police. Because of this, the first hearings of the Truth and Reconciliation Commission begun in 1995 will take place in nearby East London.

Not knowing any of this, of course, I placidly admire the passing countryside, particularly a pink grass that grows everywhere, giving the mountain slopes a magnificent pinky sheen. At dusk, when I stop for the night at the

Bungalow Hotel in Butterworth, coloured women are serving food in the dining room, the first waitresses I've seen in nearly a year.

I'm almost at Grahamstown! Setting off early the next morning I encounter small boys herding hundreds of cattle to the Kei River to drink. Camelo stops before each bunch while the lads shout and beat the animals to force them to the side of the road so I can pass. I'm lucky; I learn later that many drivers avoid travelling through the reserve because boys often hurl stones at cars that don't toss them candy or pennies. Three years later there will be a native uprising here against government control, which will end in nearly five thousand Africans being imprisoned.

I stop at Kei Road to mail letters, where three African women in long, rust-coloured skirts down to the ground stop to stare at me and grin. They walk off singing to themselves, practising modern jazz steps to go with their music.

I reach Rhodes University in time for lunch with Anne Alexander at the girls' residence where she is now house warden, then drive to the Ewers's who have kindly agreed to put me up again until my ship sails in a week, on June 2nd. In the evening, when I attend an amusing lecture by Anne called "Scorpion Tales," I catch up on news of friends from my earlier visit. Anne is still working hard on scorpions while Kit Cottrell, who is focused on ticks, has won a scholarship to study with the eminent Professor Wigglesworth at Cambridge; later, a current friend of mine in Guelph, zoologist Rita Wensler, will become friends with both Kit and Anne, who followed him to England for further research. Neville is hating his work for the government in East London and Griff, her warthog study completed, is now immersed in the intricacies of the second muscle layer of some fish.

My first task is to sell Camelo, so I run an ad in both the Grahamstown and Port Elizabeth papers offering a superior second-hand Ford Prefect for sale. Small cars don't appeal to most South Africans, though, since they love to drive long distances often over poor roads. No one phones to inquire further.

Mr. Armitage, the garage man who sold me Camelo, refuses to buy her back despite his earlier promise. "Your best bet is to drive to Port Elizabeth and sell her there," he announces.

When I tell this to Jakes, he's aghast. "That would really be stupid," he says. "Anne and I have already arranged to take you to the ship. If no one wants it, or if they find you're about to sail and have to sell immediately, they'll offer you a pittance. We'll be there on Saturday, so garages may not even be open."

I don't know what to do. It isn't fair to leave Camelo with my friends and give them the responsibility of trying to sell her. Fortunately, Jakes has arranged for a showing of my giraffe film in the main university lecture theatre, which about forty zoology students and others, all white, come to see.

"Don't forget to admire the perky car, Camelo, that enhances the movie. She's small, so she offers a good height comparison with the giraffe," he says to audience laughter as part of his introduction. "She's for sale, very reasonably, if some lucky person wants to buy her."

The movie begins with Camelo sweeping down the Drakensburg Range to the Transvaal lowveld, then manoeuvring along rough dirt tracks in the various Fleur de Lys paddocks, discovering giraffe. Fortunately, a student from Kenya who wants to buy a Rough Tough Car to drive back home is so impressed with Camelo's performance that he offers me £140 for her, which I accept.

Everyone seems to like the film. "It's amazing how much a giraffe can make itself look like a bush," one man comments. Actually, he is referring to a bush with coloured leaves, not a giraffe, but I don't correct him.

"Giraffe are amazingly well camouflaged despite their size," I say.

"It's seems quite a professional job," another comments of the film, to my pleasure.

Later in the evening Jakes, Griff, two graduate students who have read my giraffe report, and I gather at the Ewers's to discuss it. No one thinks it needs any serious revisions beyond those Griff has suggested to me.

"You must publish it in either the *Proceedings of the Zoological Society of London* or in the *Journal of Mammalogy*," she tells me, naming the two most prestigious journals for mammalogy in the world.

"Really?" I say. I'm euphoric.

"*All* scientific research should be published," she insists. "That's what science is, building on the work of other scientists."

No one has ever told me this before. Not my parents, who are both academics. Not Ian. Not even the supervisor of my master's thesis; I'd not thought beyond actually earning my degree. Did none of them take my scientific work seriously? Later I realize that publishing negative results, results that don't prove anything, is especially important for experiments dealing with animals; otherwise the same research could be carried out again and again by scientists who didn't know that it had already been proven useless, with more and more animals suffering and dying for no good reason.

The next morning, over breakfast, Griff tells me about the newest political developments in Grahamstown. "The government is putting the Group Areas Act into effect. After surveying the twenty-three thousand people living here, they've decided there will be five distinct quarters, for Europeans, Natives, Indians, Coloureds, and Chinese. Each community, in theory, will eventually be self-supporting, with their own shops and recreation facilities. No one will have any rights in any areas that they don't belong in racially."

"Wow!" I say. "I haven't even seen any Chinese here!"

"There are thirty-five of them." We look at each other in dismay.

"Each quarter has to be a hundred yards away from every other quarter," Griff goes on, rolling her eyes.

"But there are no large open areas *in* Grahamstown," I protest.

"No matter. The best native houses are going to be knocked down to make the no-man's land between the whites and the natives. Indians who own some of the best shops in town will have to move them to the Indian quarter. The Coloureds are going to stay where they are, but hundreds more people will be shifted there, which will make them terribly over-crowded."

"What a nightmare!" I gasp.

"That's not the worst," Griff says. "In the European quarters, no non-European has any real rights at all. If one of the waiters on campus loses his job or annoys white people, he can be forced out of Grahamstown entirely. He's here at the sufferance of the Europeans."

"But where would they go?"

"To the 'homelands' of each racial group, the 'Bantustans,' somewhere in the boondocks."

"Yikes!"

Much later I read that these changes aren't really about maintaining racial purity at all. They're about creating separate communities to foster enmity— divide and rule. If there is an uprising, a location can easily be sealed off by the military and its people starved if there is a general strike. The Grand Apartheid scheme will be in place by 1980, after which the only legal home of the African population will be the Bantustans. About half of all Africans will live there, while 35 percent will live illegally in the cities and 15 percent illegally in the rural "white" areas. These "illegal" people will be deported to a Bantustan only if they cause trouble. That evening I go to the bioscope to see *War and Peace*, which takes my mind off these horrors, but provides others for me to think about.

On the drive to Port Elizabeth I don't have the opportunity to feel too sorry for myself about leaving because Jakes is on a rant against the United States.

"Americans are completely materialistic—all their culture is based on money," he declares as we drive along. "A new survey says that at any one time 17 percent of Americans are reading a book, 30 percent of Canadians, and 50 percent of Brits. Their education system is so bad that it allows a child to do anything but learn. I wouldn't dream of sending one of my kids there to study. They have miserable academic standards."

"They do lots of research, though," I interject. "Many professors publish several papers a year."

"Which is why the research is so awful. Professors have to publish to get tenure and promotions—it doesn't matter if the papers are useless. How could they be good when there's little time for scholars to think and explore new ideas? They squander billions of dollars on research but produce little that's worthwhile. No other country has such a poor record of value for money spent."

"What about their discoveries during the war?" I protest.

"Done by Europeans forced to flee to America in the thirties. Their genius will die out with them. The education system is so bad that their off-spring will be no smarter than the rest of illiterate Americans."

"But..."

"And Canadians are little better." I shut up and look out the window.

Anne, Jakes, and I have lunch at an American-style restaurant in Port Elizabeth, where there's a gaudy decor, coloured lights flashing, a jukebox blaring, and many teenagers, all white, wearing fantastic clothes. I feel quite at home, but not Anne or Jakes.

"This is awful," exclaims Anne, waving her hand to include the whole ambience of garish colours and loud sound. "I can never go to America if it's like this!"

In the years ahead, Anne will do further research at Cambridge University before becoming a professor at the University of Pietermaritzburg. I try to contact her in South Africa but without success, later learning that she has been killed in a car crash. Griff died of cancer in England, and Jakes too has now died near London, ending our regular correspondence.

# 19

## *Return to England*

We say sad goodbyes at the dock beside the *Carnarvon Castle*, which sails for Cape Town in mid-afternoon. On board ship I hurry to the purser to find a letter from Ian, which I eagerly tear open. But far from recounting amusing tales from Canada, it complains that our meeting in England will be very expensive for him and travelling around the continent even more so. I'm horrified that he seems to consider me a spendthrift when in reality I'm the cheapest person I know. I write back an angry letter ending, "Oh, Ian, I am afraid this is a dreadful letter but it shows how upset I am anyway. If you answer this letter as fiercely as I am answering yours there will be no end to it or at least there will certainly be an end to us." Looking back, I imagine that I felt supremely competent and independent after my year in Africa. I obviously hated being questioned, I thought unfairly, about financial matters.

I'm dejected as I hunt for my cabin, sad to be leaving my friends, depressed to think I'll probably never see Africa and wild giraffe again, and furious with Ian. When I find my cabin, though, on the tiny table next to one bunk is an enormous bouquet of flowers sent to the ship by Martin—twelve gladioli, twelve tulips, twenty-four different shades of roses, and some unknown flowers. What an amazingly kind friend to cheer me up in this way! How much more thoughtful than Ian!

At breakfast the next day I'm placed at a table with three old ladies and a young sea cadet of sixteen.

"Would you like to change places with me?" the grandmother next to me asks after we're settled, poking me chummily in the ribs. She has a wandering eye, so I wonder if perhaps this has something to do with her comment, but I mustn't say so.

"Isn't your chair comfortable?" I ask.

All the ladies snigger at this, to my puzzlement.

"Why should we change?" I ask, completely bemused.

The women burst out laughing, so convulsed that they can hardly eat their food. Suddenly I realize that my neighbour thinks I'm about sixteen too, and is matchmaking on my behalf. She wants me to sit next to the young cadet. This makes me so angry that my hand shakes as I try to cut and eat my bacon.

That night I exercise by pacing around and around the top deck, routinely passing a little man of about forty circling in the opposite direction. Soon we are smiling shyly at each other every time we meet. After about five such encounters, he grabs me by the arm as we pass to stop me.

"I speak no English," he says. I gape at him in surprise at such an unusual opening. Then I gather myself together to be polite.

"Are you going to Cape Town?" I ask.

He shakes his head in incomprehension. He really doesn't know English. "Windhoek—German," he says, pointing to his chest. Windhoek is the capital of South West Africa (later Namibia), which had been a German colony before the First World War and so isolated that its subsequent transition as a South African mandated territory had obviously made little difference to some of the residents. Several laps later, more smiles but no words, I decide that he thought I might consider him rude for not speaking, so he felt he had to explain himself.

We dock in Cape Town for a four-day stay early the next morning. To my delight, although there's no mail from Ian, there are letters from my mother and from Oom Alex, who is feeling lonely now that I've gone and there is no one at Fleur de Lys to share his interest in wildlife. He writes that my cheerful attitude fostered and cemented hitherto unshown goodwill all over Fleur de Lys, and that my stay and our photographic efforts were one of the happiest times in his life. My mother writes with plans for our tour of England in a few week's time.

After a morning spent shopping and buying a copy of *War and Peace* to read on board ship, I attend the National Parliament in the afternoon, sitting in the Strangers' Gallery with stockings on as a sign of respect: the first time I've worn them in a year. I'm excited to be in such an august body, but for the first two hours only Nationalists speak, and always in Afrikaans. I have no idea what they're talking about. Finally a member of the United Party stands to complain about an amendment to the Group Areas Bill being discussed.

"The Bill is now so complicated, after five different amendments, that even lawyers have difficulty with it, let alone Africans," he pronounces. "This newest amendment gives the Minister of the Interior, Mr. Donges, too much

power, even though Mr. Donges says he is really only simplifying the bill."
Shouldn't the United Party be fighting the bill itself, which had so distressed
Griff, I think, rather than worrying about amendments?

"I don't want to doubt the integrity of the Minister of the Interior," the
speaker continues gravely, "but at some later date someone like Mr. Verwoerd,
the Minister of Native Affairs, might become Minister of the Interior and
that would be a different story." There is much laughter at this witticism,
which makes me wonder. Is this all a joke? Later, of course, the power of
Mr. Verwoerd as prime minister won't be a joke at all to non-Europeans in
South Africa.

The same United Party MP queries the section of the bill by which anyone can question the racial status of any other person. "This will put a pre-
mium on idle gossip, and be a terrible onus on the accused person," he states.
He's got that right.

Over the next few days, I visit the magnificent Kirstenbosch Gardens,
about ten miles away by bus, and the cosier Botanical Gardens behind the Parliament buildings. Then I join a mid-morning bus tour to the Cape of Good
Hope with fifteen other people, including a family of five emigrating to
Canada; a Cape Town couple whose son was caught in London for National
Service duty and has instead signed on to the Union Castle Steamship Line
for a five-year stint of duty; and a ladies' hairdresser who has been working
at the mines in the Copperbelt of North Rhodesia, styling, I presume, the hair
of European wives.

We drive along the coast admiring the luxury housing overlooking the
Atlantic Ocean and the Twelve Apostles, twelve mountain peaks rising behind
them. After winding through mountains and vineyards we come to addi-
tional expensive houses, resorts, and lovely beaches on the Indian Ocean.
When we stop for lunch, I sit with the Cape Town couple who tell me more
about their son; I listen so agreeably that the man insists on paying for my meal.
From here we drive south through low scrub bush supporting little wildlife
except troops of baboons, which continue to welcome tourists to this day.
When the bus stops, the baboons leap onto it and goof about; while those of
us inside peer out at them with delight, the bus driver closes his eyes and
hunches down in resignation in his seat.

At the tip of the Cape of Good Hope, we climb up a steep hill in light rain
to an old lighthouse, and gaze out to where roiling waves seem to mark the
meeting of the Atlantic and Indian Oceans. I'm thrilled to be here, at the
juncture of two such mighty bodies of water.

"The Indian Ocean on your left is a few degrees warmer than the Atlantic
Ocean," our guide announces.

"More swimmable," says the hairdresser.

"No. It's too dangerous to swim, with riptides and undertow."

As we drive north along the coast of False Bay, two black seals wrestle in the shallows. "I guess seals don't mind undertow," the hairdresser observes to no one in particular.

"There are too many seals," the guide says. "The government has to cull some each year."

We pass Simonstown, the huge naval base given a few months earlier to South Africa by Great Britain, the beach house in which Cecil Rhodes died, Cape Town University and, far above us, the white Rhodes Monument set in a lovely stretch of green lawn where it celebrates at the same time both capitalism and native oppression. Even with all this excitement I'm back on board ship in time for dinner.

After our meal I persuade a 73-year-old Canadian passenger to go with me to the movie *Alexander the Great*. Having seen *War and Peace* several days before, I'm anxious to know more about Tsar Alexander. I'm dismayed, when the movie starts and Aristotle is introduced, to realize that I have the wrong Alexander in mind.

The next day before breakfast, Oom Alex calls to wish me a final goodbye. Soon after I develop a fever for which a doctor, who tells me he lived in Winnipeg for six years, gives me codeine. This does the trick so well that I resolve not to miss the evening concert for which I've bought a ticket. Fellow passenger Chris Hollenbach, who is en route to Europe for a holiday, decides on the spur of the moment to come with me. He hails a taxi.

"Have you been to Parliament yet?" I ask him as we settle inside.

"No. I should, though, because I've just graduated as a lawyer." He has a slight accent because Afrikaans is his first language.

"The University Bill is shocking," I say. "It means that only white people can go to the best universities."

"I'm a Nationalist," Chris says, "and the law is only meant to keep the different races from mixing." We argue hotly about the bill. When we get out of the taxi Chris asks the coloured driver in Afrikaans which of us is right.

"I think the lady's right," he says.

The concert, all pieces by Elgar, is wonderful, the first I've heard for over a year. I find the *"Enigma" Variations* truly superb. At the end Chris and I have coffee at a restaurant, then walk back to the ship arguing affably about racist practices.

"Did you know that the waiters on our ship earn far more money than the Goans who work below deck?" I ask. The steward who draws my salt water bath each evening told me this when I asked him.

University procession in Cape Town protesting the apartheid
exclusion of native Africans from the best universities.

"They're white. They have a far higher standard of living."

"Of course they do, since they have much more money. And the Africans
who work in the furnace room earn even less than the Goans, though their
jobs are worse."

"It's a problem," Chris admits. "I guess the company pays only what it has
to, keeping in mind the average wage for each group."

"Weird," I say.

"It *is* weird," Chris agrees. He hadn't thought of this before.

On our last morning ashore Chris and I visit Parliament together, where
Chris points out many of the men I've been reading about in the newspapers—
Strydom, de Villiers Graaff, Donges, and le Roux. Again they speak only in
Afrikaans, so I pass the time staring at their (to me) evil faces. Pity I don't have
an evil eye myself to vanquish them. Later we walk to Adderly Street to watch
a procession of three thousand Cape Town University students and faculty,
many of them in black gowns, protesting the University Bill. The partici-
pants, mostly men and mostly white, march slowly, four abreast, unsmiling,
looking straight ahead.

I nudge Chris, "*They* certainly have problems with the bill." Chris nods
noncommittally.

We visit the Groot Kerk, a very old and simple Dutch Reformed Church,
where enclosed stalls take the place of pews. "Cape Town was settled by the
Dutch in 1652 to supply their ships going to or coming from Dutch East

Chris Hollenbach, a Nationalist friend.

Indies," Chris says. "It's far older than our other cities. The Boers were forced
north only when the British arrived and tried to abolish slavery."

"What about the Cape Coloured population? Were they slaves?" I ask.

"They're a mixture of early Boer and native blood."

"No problem with miscegenation then?" This is a dig at his horror of
white men sleeping with native women, or raping them.

"I guess not," he concedes. "They have privileges such as better schools
and living conditions that the natives lack."

After trailing over to the Supreme Court, which is closed, and attempting to visit an old doctor I had met years ago but who has, unknown to me,
since moved to America, we're back at the ship for lunch, where all is confusion preparatory to sailing. New passengers are hunting for their luggage
piled in the halls and for their cabins. Their friends hover about, anxious to
celebrate the occasion with champagne. In my room I find, from Oom Alex,
a telegram wishing me bon voyage, a gorgeous bunch of South African flowers in a basket, a box of Black Magic chocolates, and a huge box of goodies—
oranges, apples, tangerines, pears, guavas, avocado pears, granadillas, candies,
and cookies. What a magnificent man!

After a second mighty blast of the ship's horn to ensure all visitors have gone
ashore, the *Carnarvon Castle* is slowly pulled by tugboats out of the harbour, we
passengers lining the rails to wave at those on the quay growing smaller and less
distinct. A band plays lively music but the elderly woman next to me shakes with
sobs, her face contorted with grief at leaving her home or loved ones.

My roommates are Mrs. Plastow, aged seventy-five, and Miss Davies, aged sixty, both Englishwomen with disabilities. Needless to say, I volunteer for the upper berth. Mrs. Plastow, a wonderful storyteller, lived for some years in backwoods Canada. One day when she was melting lard on her wood stove, the pan tipped over and set her clothes on fire. She rushed outdoors and rolled in the snow to put out the flames, then walked a long distance in a blizzard to get help. On the way her burns became frost-bitten and she had to spend two years in hospital; when she emerged, one leg had been amputated and her face was partially paralyzed. She now props her artificial leg against her berth when she's in bed, hopping down the hall to the bathroom if she has to get up in the night.

Mrs. Plastow finds the boat dull; she's been staying with her daughter in Cape Town, getting up each morning at five o'clock to cook meals for the family of nine, which kept her busy. She plans to amuse herself on the ship by reading.

"I'll read the telephone book if I have to," she tells me. "I love to read." I lend her André Gide's *Journal,* which she enjoys.

Miss Davies, who has been visiting friends in Southern Rhodesia, also has part of her face paralyzed, in her case from shell-shock.

"I gave up eating fish during the war," she announces to our cabin. "So many men drowned at sea that I knew eating fish would be like cannibalism." Neither Mrs. Plastow nor I had thought of this aspect of the war disaster. History repeats itself. In Sri Lanka today, after the tsunami of December 2004, people no longer eat fish for the same reason, which has devastated a major part of their economy.

In the dining room I share a table with a black woman, Mrs. Susan Wagi, and four elderly South African women, three of them huge and obsessed with food, which is a major topic of our general conversation.

"The potatoes are always fried and always greasy," one complains.

"We had pork chops yesterday," another comments on noticing them again on the menu.

"Does she like our food?" the third woman asks me, motioning to Susan. She seems to think that Susan doesn't speak English, or can't hear.

"Do you like this food?" I ask Susan as if I really am an interpreter, smirking at her because of this.

"I would if it were better cooked," replies Susan, a midwife who has been in South Africa for the birth of her first child, Vuyo, now a year old. She is returning to London to rejoin her husband, a Tanganyikan who lectures in mathematics at the University of London.

"Look at what the waiter is doing, again!" Susan says to me later. We both watch in revulsion as he wipes our used knives and forks off with a damp cloth and returns them to his drawer so he won't have to go to the kitchen to wash them properly.

"That other waiter wipes the water glasses that way too," I remark.

In the mornings, although I'm suffering from periodic headaches, I edit my giraffe article, write in my journal, sort my slides and newspaper clippings, and read *War and Peace*. Each afternoon I spend time with Susan and Vuyo. Susan has agreed to let me write a brief story of her life, so we discuss this for several hours. Her situation is difficult aboard ship, even though there is ostensibly no racial discrimination.

"When I go into the lavatory I sometimes flush both toilets," she tells me, laughing. "When other women come in, they don't know which stall to avoid."

We play with Vuyo, who giggles when I carry her about on my hip. "I've been hired as a nanny," I say cheerfully to those people I recognize, watching to see how they'll react. Most look baffled. Susan and Vuyo become the focus of my social life. Sometimes Chris Hollenbach joins us to chat; he tries to be friendly with Susan, but as a Nationalist he isn't sure how to do this.

"I don't know what people think of me, talking to a black girl," he confesses to me. He helps with Vuyo's baby carriage when Susan goes up or down stairs, Susan makes a point of telling me.

"He always calls me Mrs. Wagi though, never Susan," she notes. Chris wins both a bridge prize and a chess prize for competitions on board, and reads my copy of Dostoyevsky's *The Idiot*, which he loves. We continue to argue about the racist policies of the Nationalists, especially the ban on miscegenation.

"Did you know that if a white man and black woman sleep together, the woman is legally blamed far more than the man?"

"So?" asks Chris.

"If there's a crime committed, surely they must be equally at fault," I insist.

Chris, who had never thought of it this way before, looks puzzled. "It does seem odd," he admits. (Several years later he writes me to say that, thanks in part to my advocacy, he is no longer a Nationalist and is hoping to start a more liberal party. He had enjoyed travelling in Europe in part because he was able, for the first time in his life, to chat with Africans and Indians as equal individuals.)

In contrast to Chris, Eddy Luck, a journalist and photographer from the Copper Belt who is setting off to travel around the world, is greatly favoured by Susan, with whom he laughs and jokes.

Susan Wagi and her daughter Vuyo aboard the Carnarvon Castle.

"I'll take cigarettes on shore for you if you want, Susan," he tells her, to her delight, because her husband smokes and this way she will save money by avoiding duty taxes. (When Chris hears about this he buys her a box of fifty cigarettes, which he gives me to deliver to her.) Eddy owns a typewriter so I chat with him at length, hoping to borrow it so I can retype my scientific giraffe paper. This works sometimes. However, he also knows Sir Roy Welensky, prime minister of the Federation of the Rhodesias and Nyasaland, who is travelling to Britain in the first-class section of the ship; Welensky also needs to borrow it now and then, so I'm second in line.

"I had drinks again with that coloured chap," Eddy often says to me when we meet, referring to another passenger. And "Did you know that Chris Hollenbach is a bloody Nationalist?"

At a talent contest one evening Eddy, who is a large, six-foot man, dons a mini tutu over his bathing suit and pirouettes to ballet music, which brings long rounds of applause.

"Will you marry me?" he asks after drinks celebrating his performance.

"I don't even know you," I laugh.

"I'm dependable as Star," he contends; he had read part of my giraffe report when he lent me his typewriter.

"But Star is hardly a model of dependability!"

Toward the end of the trip my headaches become more frequent. I go to the ship's doctor, wondering if he'll be as forward as the doctor in Dar es Salaam. He isn't.

"Take a couple of aspirin," he advises, rolling his eyes at my non-problem, then nodding sagely. "You'll be all right in a day or two."

I spend some time in a lounge chair on deck, closing my eyes and mulling over the past year. All my encounters with giraffe were as wonderful as I had imagined they would be. My reflections on the people I met in Africa, by contrast, baffle me. Why was Mr. Matthew, Oom Alex, so kind to me yet so negative toward natives? Why this dual personality? It seems to me that he must have different compartments in his brain, a large one of neutral or positive thoughts for Europeans and another with negative opinions of natives. Somehow these two boxes don't speak to each other. How else to explain how he gave me the possibility to prove myself as a zoologist, but didn't think natives deserved a chance to be properly educated? Maybe his brain holds a third box for women, too, but a more porous one since it doesn't apply to all of us. He was wonderful to me, but I recall how, at Victoria Falls, he refused to be civil to a woman who merely wanted to chat with him.

Now, writing up this memoir fifty years later, I ask myself, did I too have separate brain boxes? To be honest, I have to answer yes, although the thought that I was also racist horrifies me. But why else would I not have arranged somehow to help Mokkies and the others in some real way? Maybe I could have taught Bella and Enoch to read and write. I didn't have much money while in Africa, but I could surely have planned to send some from Canada when I returned there. I may have idealized Africans in general, but I obviously didn't really think the workers at Fleur de Lys were as smart as me, or as important. I knew that Josiah Chinamano was my equal, but then he wore a shirt and tie, read the same books I did, and spoke English. So too was Susan Wagi my equal. I obviously had one brain box for Josiah, Susan, and idealized Africans, and one for nondescript Africans in general.

Later a psychology student who listened to my theory of brain boxes suggested a new idea to me. He said that it related to cognitive dissonance, first described in 1957, the year I was in Africa. According to this theory, because people want their attitudes, thoughts, and behaviours to be consistent with one another, they subconsciously organize their mindsets to make this happen. Behaviour is especially hard to change, so South Africans such as Mr. Matthew, who use Africans as cheap labour, persuade themselves that racism is acceptable because natives are stupid and need the work. Up until the end of apartheid, virtually all white South Africans thought this way, their children in turn naturally accepting this attitude as well; indeed, any white South African who treated blacks as equals was vilified by their peers. Slowly, with the abolishment of apartheid and the first free elections in 1994, the chasm between blacks and whites seems to be narrowing.

Luckily, today I'm a different person with a less idealistic view of race, a word I now know is meaningless given the mixing of bloodlines over the centuries. I've always known that blacks and whites should be treated the same way, but in the past I tended to put individual blacks on a pedestal, the same pedestal from which I try, as a feminist, to liberate women. At the University of Waterloo, I've been thrilled to work in the past with black students merely because they were black. I only came to my senses on a Toronto bus in 1987, on the way to a conference. A fellow attendee, judging by her name tag, a black woman in a pink suit, looked so marvellous that I almost said to her, a stranger, "That suit looks great on you!" But I hesitated. Would I have said something like that to an attractive white woman I didn't know? Of course not. Like a shot of adrenalin, I finally realized once and for all how ridiculous my mindset had been. Now I'm free to find some blacks endless fun, some interesting, and some boring, just like people everywhere.

We arrive in Southampton early in the morning of June 20, 1957, dawn lightening the sky in the east. I stand with Chris at the railing of the ship, watching men bustling about the dock, tying up the ship. He's amazed that the dockhands are white—he's never seen a white man doing manual labour before—while I'm grieving because my dream of a lifetime is over at age twenty-four. I fear that I will never again visit the giraffe in Africa, and I never have.

# EPILOGUE

I feel sick for days after settling into a cheap hotel in London, although better when I totter over to Canada House in Trafalgar Square to collect my mail and find an airmail letter from Ian apologizing for his letter that upset me. He hadn't meant to sound so negative, he insists, but merely to suggest that we should save some money for a rainy day. About the letter he writes, "My only advice and hope is that you take it and tear it into little shreds or burn it but above all—try to forget it. I am really sorry for it." And this is what I must have done: of the scores of letters that passed between us during the year (mostly from me), only that one is missing. Ian says he plans to come to England in late August when we can be married. Marriage seems like a good idea, even though he doesn't add any xxxs when he signs off "Love" at the end of his letter. Now that my dream of a lifetime is over, I don't have any plans for the rest of my life.

When I go to meet my mother's plane at Heathrow Airport a week later, I'm flushed red with a high temperature. My mother hurries me back to the hotel where she phones a doctor, as a mother would. He comes quickly and, under the new National Health scheme, gives me a free injection of an antibiotic. When I go to the nearby Royal Free Hospital (I like the "Free" in the name) for a checkup a week later, I'm put in a wheelchair and rolled into a lecture room of steeply banked seats before a large class of medical students.

"This patient has come from Canada to England by way of South Africa," the professor, referring to his notes, announces to the class as the students stare at me. "How long did that take?" he asks. They wait to hear my answer but I'm left speechless. What should I say? How could South Africa be on the way

to England from Canada? Should I object to his question? Try to straighten
out his geography? Answer that it took me a year?

"This patient suffers from aphasia," the professor finally declares to
explain my silence. I'm wheeled out the door as a new patient is wheeled in.

I arrange at St. Pancras Townhall to sign in as a resident of the borough
so I'll be able to buy a wedding licence there after fifteen days. Then my
mother and I rent a car and drive to Scotland, on the way nearly killing a man
who strides confidently in front of the car on a crosswalk marked with white
stripes. I hadn't even slowed down, not knowing there was such a thing as a
zebra crossing for pedestrians.

In the Vale of Glencoe we approach a brand new Land Rover sporting a
Canadian logo and small, meticulously scenes painted on it, parked by the side
of the road. Inside are my pals Bris Foster and Bob Bateman, making tea. I park
the car behind their vehicle and tiptoe to its back door.

"Small world," I say.

They turn to stare, recognize me, and fall about in welcome. My mother
is as delighted as I am to share crumpets and adventures.

On the way south again, we start early to worship at Abbotsford, the
home of Sir Walter Scott, one of my mother's heroes.

"I've wanted to visit Abbotsford my whole life," she enthuses.

When we reach the entrance, however, we learn that the estate isn't open
to tourists for another five hours. We decide not to wait, as if five hours makes
much difference in a lifetime.

The week before Ian is to arrive by plane, my mother decides that her back
is too sore from sleeping on soft hotel beds for her to remain any longer in
England. She books a flight home, on the way sitting next to an excited woman
flying to Winnipeg to see her own daughter married. Mother feels guilty she
has decided not to stay for mine.

When Ian arrives on a Sunday, I recognize him despite my fears. We
exchange a magnificent hug. I've been ambivalent about getting married, but
now that he has his arms around me it seems like a terrific idea. Even so,
when we visit a jeweller next day to buy a wedding ring for me (he refuses ever
to wear one), I insist on paying for it myself so that I don't feel I am being
"bought" or am losing my independence. The wedding licence indicates that
we are free to marry the next Wednesday, but Ian feels that Thursday will be
better because we have so much to do to get ready. I'm amazed at this delay.
However, Thursday works well too and is the beginning of thirty-five years
of a wonderful marriage, in which Ian proves far more willing to spend money

than I, despite his earlier worry. Our marriage will be cut short in 1993 by his death from a massive heart attack, following one of our thrice-weekly games of mixed-doubles tennis.

When we return to Canada, I show our giraffe film to a member of the Canadian Broadcasting Company in Ottawa. "It's a good film," the photographer says, "but no television watcher would sit still to see more than one or two minutes of giraffe at the most." Luckily he is, in time, proved wrong because many years later, Public Broadcasting WNED produces an entire documentary on giraffe, for which it purchases some of our footage of males sparring.

In 1959 we move to Waterloo, Ontario, where Ian is hired as a professor of physics at the new University of Waterloo. There we have three children, Hugh, Ian, and Mary. While caring for these children, I earn my PhD in biology in 1967 from this university, writing a thesis on the gaits of large mammals, including those of giraffe, obtained by tracing leg positions from successive frames of my 16 mm movie. I extend my research on gaits in the summers of 1972 and 1973, taking films, to be analyzed later, of walking, pacing, and galloping camels in Mauritania. I write up these adventures in a trade book, *Camel Quest: Summer Research on the Saharan Camel* (1978) and in an academic tome, *The Camel: Its Ecology, Behavior and Relationship with Man* (1981) co-authored with Hilde Gauthier-Pilters.

Early on, hoping to make new friends, I tell a few acquaintances about my giraffe studies but this proves to be a mistake. Two of them think I'm lying and tell me later with surprise, after they've checked out my stories, that they've found they really are true. From that time on I no longer feel free to talk about my connection with giraffe.

Ten years after my stay in Africa, I spend an afternoon with Griff at the London Zoo en route with my family to a sabbatical year in Australia. We glance at the caged animals we pass, but mostly discuss in detail our respective research projects, mine the book on giraffe I'm writing and hers her newly published *Ethology of Mammals*. It's the first time in ten years I've been able to discuss my work with someone who understands and shares my interests. Euphoria!

"How did you manage to keep on being a scientist with small kids?" I ask her, conscious of Ian and my three young children straggling along behind us.

"It helps to have servants," she remarks.

When son Ian screams on being stung on his lip by a wasp, she rolls her eyes in exasperation as I turn to comfort him. It's difficult being both a scientist and a mother.

The next day my family and I visit the Matthews, who are retired and living in London. I continue to correspond with Oom Alex until he dies of old age, in the late 1990s.

Despite a love of university teaching on a part-time basis and success at it, I'm unable to land a permanent position as a scientist at any of the area universities; in the 1960s and 1970s, they rarely hired women on regular appointments.

"I'd never give tenure to a married woman," a science dean tells me in 1972, implying that since such a woman has a husband to support her, what is her problem? She can work part-time and do research on her own if she wants to.

So this is what I do. While employed part-time in the Independent Studies Program at the University of Waterloo I continue investigating various facets of giraffe using my own money and time, publishing many articles and co-authoring a book on giraffe. More recently I've carried out research on feminist issues (to fight against sexual discrimination, especially at universities), on homosexuality in animals (because of the behaviour I observed in giraffe and prejudice against my gay and lesbian students), on sociobiology, and on animal rights.

# APPENDIX 1

## List of Scientific Publications Involving Giraffe by the Author

1958 The behaviour of the giraffe, *Giraffa camelopardalis*, in the eastern Transvaal. *Proceedings of the Zoological Society of London* 131: 245-78. (By A.C. Innis)

1959 Food preferences of the giraffe. *Proceedings of the Zoological Society of London* 135: 640-42.

1960 Gaits of the giraffe and okapi. *Journal of Mammalogy* 41: 282.

1962 The role of the neck in the movements of the giraffe. *Journal of Mammalogy* 43: 88-97.

1962 The subspeciation of the giraffe. *Journal of Mammalogy* 43: 550-52.

1962 The distribution of the giraffe in Africa. *Mammalia* 26: 497-505.

1962 Giraffe movement and the neck. *Natural History* 71(7): 44-51.

1963 A French giraffe. *Frontiers* 27(4): 115-17,129.

1965 Sexual differences in giraffe skulls. *Mammalia* 29: 610-12.

1968 The walking gaits of some species of Pecora. *Journal of Zoology* 155: 103-10. With A. de Vos.

1968 The fast gaits of some Pecoran species. *Journal of Zoology* 155: 499-506. With A. de Vos.

1968 External features of giraffe. *Mammalia* 32: 657-69.

1970 Tactile encounters in a herd of captive giraffe. *Journal of Mammalogy* 51: 279-87.

1970 Preferred environmental temperatures of some captive mammals. *International Zoo Yearbook* 10: 127-30.

1971 The scientific name of the reticulated giraffe; proposed rejection of *Giraffa camelopardalis australis* Rhoads, 1896. *Bulletin of Zoological Nomenclature* 28: 100-101. With W.F.H. Ansell.

1971 *Giraffa camelopardalis.* Mammalian Species Paper No. 5: 1-8. New York: American Museum of Natural History Series.

1972    Notes on the biology of the giraffe. *East African Wildlife Journal* 10: 1-16. With
        J.B. Foster.

1973    Gaits in mammals. *Mammal Review* 3: 135-54.

1976    *The Giraffe: Its Biology, Behavior, and Ecology.* New York: Van Nostrand Reinhold.
        With J.B. Foster.

1979    The walk of large quadrupedal mammals. *Canadian Journal of Zoology* 57: 1157-
        63.

1982    Supplementary material prepared for the updated reprint edition of *The Giraffe*,
        published by Krieger Publishing, Florida.

1983    Homosexual behavior and female-male mounting in mammals—a first survey.
        *Mammal Review* 14: 155-85.

Entries on "Giraffe" for the *Elsevier's Animal Encyclopedia* (1970), *World Book* (1987), and
*Book of Knowledge* (1988).

# APPENDIX 2

## Pioneer Behavioural Research on African Mammals

| Date | Species | Scientist | Nationality | Country |
|------|---------|-----------|-------------|---------|
| 1956 | Giraffe | Anne Innis (Dagg) | Canadian | South Africa |
| 1957 | Kob antelope | Helmut Buechner | German | Uganda |
| 1958 | Baboon | K.R.L. Hall | British | South Africa |
| 1959 | Gorilla | George Schaller | American | Congo |
| 1959 | Wildebeest | Lee/(Martha) Talbot | American | Tanganyika |
| 1960 | Chimpanzee | Jane Goodall | British | Tanganyika |
| 1962 | Zebras | Hans Klingel | German | Tanzania |
| 1963 | Impala | Rudolf Schenkel | German | Kenya |
| 1964 | Spotted hyena | Hans/Ute Kruuk | Dutch | Tanzania |
| 1964 | Black rhino | John Goddard | Canadian | Tanzania |
| 1965 | Afr. elephant | Iain Douglas-Hamilton | British | Tanzania |
| 1966 | Afr. buffalo | A.R.E. Sinclair | Rhodesias | Tanzania |
| 1966 | Cheetah | Randall Eaton | American | Kenya |
| 1966 | Lion | George Schaller | American | Tanzania |
| 1967 | Dikdik | Ursula/Herbert Hendrichs | German | Tanzania |
| 1968 | Gerenuk/lesser kudu | Walter Leuthold | German | Tanzania |

Data largely from Cynthia Moss, *Portraits in the Wild* (Boston: Houghton Mifflin, 1975).

# SELECTED READINGS

Baxter, Joan. 2000. *A Serious Pair of Shoes: An African Journal.* East Lawrencetown, NS: Pottersfield Press.

Dugard, Martin. 2003. *Into Africa: The Epic Adventures of Stanley and Livingstone.* New York: Doubleday

Duke, Lynne. 2003. *Mandela, Mobutu, and Me.* New York: Doubleday.

Finnegan, William. 1986. *Crossing the Line: A Year in the Land of Apartheid.* New York: Harper and Row.

Heminway, John. 1990. *African Journeys: A Personal Guidebook.* New York: Warner Books.

Kapuscinski, Ryszard. 2002. *The Shadow of the Sun.* Toronto: Random House.

Mallows, Wilfrid. 1984. *The Mystery of the Great Zimbabwe.* New York: W.W. Norton.

Pifer, Drury. 1994. *Innocents in Africa: An American Family's Story.* New York: Harcourt Brace.

Ridgeway, Rick. 1998. *The Shadow of Kilimanjaro: On Foot across East Africa.* New York: Henry Holt.

Shaffer, Tanya. 2003. *Somebody's Heart Is Burning: A Woman Wanderer in Africa.* New York: Vintage Books.

Theroux, Paul. 2003. *Dark Star Safari: Overland from Cairo to Cape Town.* Boston: Houghton Mifflin.

Turner, Myles. 1987. *My Serengeti Years.* Edited by Brian Jackman. London: Elm Tree Books.

Van der Post, Laurens. *Venture to the Interior.* London: Hogarth Press.

# GLOSSARY

| | |
|---|---|
| *ayah* | a non-white woman servant who cares for young children |
| *copra* | dried coconut meat |
| *dhow* | a wooden boat with sails used for centuries by Arabs |
| *duka* | a tiny store run by Indians |
| *kaffir* | an insulting word for native Africans, comparable to "nigger" |
| *koppie* | a steep rocky outcropping |
| *kraal* | an enclosure of branches, boards, or fencing for cattle |
| *rondavel* | a round thatched hut with walls of mud and wattle |
| *shamba* | a small holding with a house and farm plot |

# Life Writing Series

In the **Life Writing Series**, Wilfrid Laurier University Press publishes life writing and new life-writing criticism in order to promote autobiographical accounts, diaries, letters, and testimonials written and/or told by women and men whose political, literary, or philosophical purposes are central to their lives. **Life Writing** features the accounts of ordinary people, written in English, or translated into English from French or the languages of the First Nations or from any of the languages of immigration to Canada. **Life Writing** will also publish original theoretical investigations about life writing, as long as they are not limited to one author or text.

Priority is given to manuscripts that provide access to those voices that have not traditionally had access to the publication process.

Manuscripts of social, cultural, and historical interest that are considered for the series, but are not published, are maintained in the **Life Writing Archive** of Wilfrid Laurier University Library.

*Series Editor*
Marlene Kadar
Humanities Division, York University

*Manuscripts to be sent to*
Brian Henderson, Director
Wilfrid Laurier University Press
75 University Avenue West
Waterloo, Ontario, Canada  N2L 3C5

# Books in the Life Writing Series
## Published by Wilfrid Laurier University Press

CPSIA information can be obtained
at www.ICGtesting.com
Printed in the USA
BVHW051646090320
574528BV00014B/1408

9 780889 204638